Penguin Education – Unesco

Planet in Peril?
R. F. Dasmann

Planet in Peril?
Man and the Biosphere Today

R. F. Dasmann

Penguin Books – Unesco

Penguin Books Ltd, Harmondsworth,
Middlesex, England
Penguin Books Australia Ltd,
Ringwood, Victoria, Australia
United Nations Educational, Scientific
and Cultural Organization,
Place de Fontenoy, 75 Paris 7^e, France

First published 1972
Copyright © Unesco, 1972
The author is responsible for the choice
and presentation of the facts presented
and for the opinions expressed
Made and printed in Great Britain by
C. Nicholls & Company Ltd
Set in Monotype Plantin

Contents

Preface

A few days before his death, Adlai Stevenson, the United States representative at the United Nations, addressing that organization's Economic and Social Council in Geneva, said: 'We travel together, passengers on a little spaceship, dependent on its vulnerable supplies of air and soil; all committed for our safety to its security and peace, preserved from annihilation only by the care, the work, and I will say, the love we give our fragile craft.' This was in 1965. Six years later, the notion of 'spaceship Earth' has become a commonplace all over the world, while governments, international organizations and scientific institutions are mobilizing their intellectual and financial potential to face the impending dangers.

Only recently have the leaders of the world and the average citizen become aware of the so-called 'environmental crisis', though this had been forecast by scientists long ago.

Among the specialized agencies of the United Nations system, Unesco was the first to realize the dangers inherent in the uncontrolled growth of the human population and in the deterioration of the natural environment. As long ago as 1948, in his last annual report to the General Conference of Unesco as Director-General of the organization, Julian Huxley stated:

A wholly new world situation has arisen through the rocket-like rise of world population in the last two hundred years. The nineteenth century saw the figure of one thousand million reached for the first time, the twentieth that of two thousand million. The increase at the present time is unprecedented in absolute amount. The net increase of population in the world today is just about two every three seconds – over 55 000 a day, about 20 000 000 a year, well over 500 000 000 in a generation. The three thousand million mark will certainly be topped early in the twenty-first century, whatever checks in increase are attempted.

Unreflecting optimists will say that migration will bring new land into use, that new techniques will make old land more productive and that science will discover new resources, new methods, and even new synthetic sources for manufacturing food. That may be so. But even if it were so, it would not change the essential issue, but merely alter the time at which it will become acute. The final and inescapable fact is that resources are limited, in the last resort, by the space on the land's surface. On the other hand, the human multiplicative faculty continues to exert itself, and the absolute increase of population, though already greater than at any previous time, is still showing acceleration. Somehow or other population must be balanced against resources or civilization will perish. War is a less grave and less inevitable threat to civilization than is population increase.

During the twenty-three intervening years, the situation has worsened: the human population has already passed the 3 500 000 000 mark, and it is being increased at the rate of more than two children every second. The uncontrolled growth of our species acts as a sort of cancer on the biosphere, that thin layer of soil, air and water that covers the surface of the Earth and within which all life exists. Within this layer a complex series of processes and interactions has been taking place for millions of years, and this has led to the emergence of modern man. The role of man in these processes and interactions has been relatively small until recently. But in the last century, and more particularly in the last decades, the impact of man on the biosphere has grown at an exponential rate. It has become global in scope and has reached a point where the entire system which has made life – and particularly human life – possible is basically affected.

It is for this reason that Unesco, which since its earliest days has served as a catalyst in the development of ecological studies throughout the world, has just launched a new long-term scientific programme on 'Man and the Biosphere'. The object of this international research programme is to provide baseline data for all those who will have to take decisions aimed at reversing the dangerous trend towards the destruction of plant, animal and human life on our planet.

A task of this magnitude cannot be carried through without the active support of public opinion. The purpose of this book is therefore to inform public opinion, to make it aware of the urgency of the work to be done, to show it the sacrifices that will have to be accepted.

It describes the present state of our knowledge on a subject which is all too often disregarded – that of interactions between man and his environment. It sets out clearly and forthrightly the crucial problems facing the whole of mankind today, and urges us to take informed and rational decisions regarding the pressing and unavoidable choices before us. May this appeal be heeded: for it is the survival of the human race that is at stake.

A. Buzzati-Traverso
Assistant Director-General for Sciences, Unesco

Chapter 1
The Environmental Crisis

When rockets from the Soviet Union and the United States travelled beyond the limits of the Earth and on beyond the moon to explore and report back on the state of our sister planets, Mars and Venus, they shattered what had been, for some, a fond illusion. The illusion was that perhaps we were not alone, that our neighbouring worlds may be inhabited by other beings, or inhabitable by man. But the photographs and instrument readings sent back to us from the planets most like Earth held no hope. One world was observed that is like the Earth might have been over three thousand million years ago – heated beyond the tolerance of life; and the other was not unlike the lifeless moon or perhaps like Earth may someday become if all life vanishes. We are alone in our corner of the universe, travellers on our beautiful, fragile, blue planet, shielded only by its atmosphere from lethal, life-destroying radiation from our own sun and from distant stars – radiation that would otherwise leave our lands as barren as the moon. Somewhere out among the stars there may be other planets like ours, but they are beyond our knowing at distances so great that generations of men would live and die before they could be reached by the fastest spaceship we can imagine.

Not very long ago and quite without the knowledge of most people, the human race was given power beyond any normal, rational comprehension – power that our ancestors, perhaps wisely, would have reserved only for the use of gods. The power came gradually – first fire, hundreds of thousands of years ago when men were few and could do little damage to the planet. But it was a force no other animal could use and man was to use it to change the lands that he inhabited. Next came the power of machines – simple at first and powered by hand, then growing more complex and finally to be fuelled by fire and water. Only recently man extended his sources of power when he learnt to tap the great reserves of ancient fuels – the coals, petroleum and natural gases, stored in the Earth over millions of years. Still more recently he discovered the secret of transmitting power over distances – through wires in the form of electricity and then without wires in radio waves. Using the materials of the Earth – its fuels, water, minerals – mankind could transform the surface of the Earth to a degree previously achieved only by the great forces of the planet itself – earthquakes and tidal waves, hurricanes and volcanoes. The human race could, and did, use the forces available to it to enrich life and enhance the existence of mankind. But the power that was achieved was not always used wisely. Man could, and did, also use these forces to make great areas of land and water barren – to create wastelands and to poison waters. Nevertheless, while men were still attempting to achieve wisdom in the use of the power already available, they were given the power that fires the sun. Held within the smallest components of matter, the forces that tie the atom together were beyond imagining – until the day when they were partially released by the first atomic explosion.

We are the custodians of the only planet friendly to life in the known universe. We have the power to make it lifeless. In our hands, today, rests the future of all known living things. If we use our power wisely, life will continue to thrive on this blue planet and all of mankind may look forward to the future. But it is evident wherever we go on the globe that we are not moving fast enough towards the necessary wisdom. This is the basis of our environmental problems.

To a man unaided by machines or the fuels that power them, the Earth can seem endless. On foot he can travel perhaps forty or fifty kilometres in a day, but the circumference of the Earth is forty thousand kilometres and its diameter from pole to pole is nearly thirteen thousand. Yet, the great bulk of the planet is without life and beyond the reach of man. Only the outer surface of the globe – a film of air and water and soil that has been compared, in relation to the total size of the Earth, to the skin of an apple – is capable of supporting living things. This surface layer in which all life exists is known as the *biosphere*. It includes all of the oceans, down to a depth of less than eleven kilometres in the deepest part of the Pacific, the surfaces of the continents and islands up to a height of less than nine kilometres at the top of Mount Everest, and the lower levels of the atmosphere – no more than twenty-four kilometres from the deepest point at which life exists to the highest point where it can, without protection, be carried and survive in currents of air. All parts of the biosphere are now within the reach of man – he can explore the deep oceans and the farthest reaches of the sky. All parts of the biosphere have been affected by man – lead from motor car exhausts in the great cities has been found in the polar ice caps; pesticides sprayed on farmlands in temperate zones are carried in air currents to contaminate the food of polar bears in the arctic and seals in the antarctic. Once there were known areas inhabited by man, and beyond their frontiers, great stretches of the Earth where no man had been. Today, within the biosphere, there are no more frontiers – there is no country beyond the known boundaries – the limits of the Earth have been reached.

Not only is the entire biosphere now accessible to man, but new networks of transport and communication now bring all parts of it within easy reach. With the fastest planes the farthest part of the Earth's surface can be reached in little more than half a day. Radio-communication now makes it possible for people everywhere to know, within minutes, what has happened in any one place. Great events could once transpire on Earth without most people knowing of them. Now all can know and be affected by any catastrophe, any conflict. Once wealthy nations could enjoy their luxuries without the envy of peoples elsewhere. Now all want a share of the riches that can be produced from the Earth. But all also share the perils that arise from man's use of the biosphere. Once it could be said that pollution was a local problem. Today pollution is a global problem, and none can escape its consequences. Misuse of the land was once the worry of only those who occupied that land. Now all can be affected by those actions that lower the capacity of the biosphere to support living things.

Most of what man had done since the human species first appeared on Earth has been intended to make his conditions of life more agreeable. He has sought to make the Earth produce more of the things that he required – food and fibre, green plants or animal products – and he has sought to obtain more of the non-living things that were stored in the Earth but suited to his needs – metals and minerals, building materials and fuels. If we make an exception of the actions of mankind in times of violence or war, the activities of the human race have been directed towards making the Earth a better place to live in. Over large areas of the planet, these activities have been successful – conditions as they existed before man appeared on the scene have been improved from man's point of view, the environment for people has been improved. Why then have these activities led to environmental problems of such severity as to require the attention of all nations of the world? The answer can be found in four factors:

1. The increase in human populations to levels beyond any earlier expectations, and their continuing rapid increase;

2. Man's failure to control the new power and technology that is available to him and to use it in ways that would not damage his environment;

3. Man's failure to control his use of the land in order to keep this use within the capacity of the biosphere to provide for its continuing support;

and finally, although this is involved in all of the other three,

4. Ignorance or inadequate information about the environment and the ecological rules that govern its continued health and stability.

Although all of these factors operate throughout the biosphere, in any one country or region each of them may be of greater or lesser importance. It is worthwhile, therefore, to explore them in some greater depth.

The numbers of people

In 1970 more than 3 600 000 000 people inhabited the biosphere and they were increasing at a rate approaching 2 per cent per year. This meant that more than 72 000 000 additional people were being added to the world population each year – enough people each year to form a large and important new nation. This is a simple statement, yet it is beyond historical precedent.

Nothing like it has ever happened on Earth before. Furthermore, each year, because the total population is larger, the number of new people added also increases. When the population reaches 4 000 000 000, which will be soon if the growth rate continues at 2 per cent, 80 000 000 new people will be added each year. Each one of these people hopes or expects to be provided with

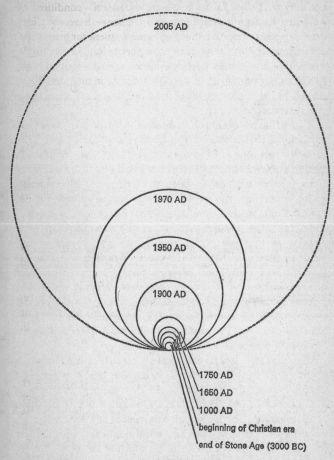

Figure 1 The size of the world population at various dates

food and clothing and shelter, and have access to the various necessities that go with civilization and to the luxuries that civilizations produce. Yet this increase occurs and this population exists in a limited biosphere, on the thin skin of a small planet.

How new and how unprecedented this population increase is can best be observed by contemplating the situation on Earth not very long ago. In 1650, when new power and technology were first becoming available to a few people in Europe, there were only five hundred million people on Earth. This represented perhaps one million years or more of population growth, during which time the human species evolved, spread over the Earth, settled in large numbers in its more favourable places, and proceeded to shape their environments to make them more habitable and productive.

During the next century, because of their gains in power and technology, the populations of Europe and parts of Asia increased. Over much of the world, however, there was either no gain in population or, in some areas, because of the impact of European colonization, a population decrease. Nevertheless, within a century, the world's population had grown by two hundred million (less than three years' increase at 1970 rates). After 1750, the population of all of the continents began to increase as the new power and technology brought greater yields of food and more of the other necessities that man requires, and as control over disease and new means for sanitation became widespread. By 1900 there were slightly more than 1 500 000 000 people. In the next half-century, however, nearly 1 000 000 000 more were added. Then within the ten years from 1950 to 1960, as many more people were added to the Earth's population as had existed on Earth in 1650.

Why has this population increase occurred? The answers are easy to find. Population growth in any one region takes place when births and immigration exceed the total losses from deaths and emigration. For the Earth as a whole there is no emigration or immigration, so the equation is simplified: growth occurs when births exceed deaths. The world population increase, for the most part, has occurred not because the birth-rate has increased, but because the death-rate has decreased. More people are living for longer periods of time. This has been a matter of great self-congratulation for humanity. It represents major gains in our ability to make the Earth produce more of the things that we require, and particularly represents gains in our ability to control the diseases and other misfortunes that caused people to die young. However, with the power to control death should have come a willingness to limit birth. If this had occurred, more people would be living for longer under conditions of increasing affluence. As it is, in many parts of the world, more people are living longer under conditions of unameliorated privation.

The picture is not simple and population growth does not occur at the same rate everywhere. Some populations are either not growing, or grow very slowly. These are either people not yet reached by the benefits of increased food, better sanitation and medical care – people that still suffer from a high death-rate – or else people who have been willing to control their birth-rate. The latter group includes both primitive people who have recognized the limits of their environment, and technologically advanced people who seek to maintain or increase their personal well-being by restricting their numbers.

The facts about world population require neither debate nor emotion. They exist. The conclusions to be drawn from these facts vary between and within nations. There are certain conclusions that are inescapable, however. One is that, if existing rates continue, there will be approximately twice as many people in the year 2005 as there were in 1970. One must ask whether, since we have proved incapable of providing adequately for the 3 600 000 000 that existed in 1970, we have the right to assume that we can provide adequately for twice as many within the space of thirty-five years. By what miracles will this be accomplished? And one must recognize that we do not expect the world to end in AD 2000; if we cannot solve the problems of population growth now, how do we expect to solve them then?

There are many parts of the world with relatively few people, for example Chad and Mali together cover about two and a half million square kilometres and had, in 1970, fewer than nine million people, less than are jammed into New York City. But most of their land cannot be made economically productive and most of their people do not live beyond a mere subsistence level.

There are countries with great numbers of people and a high standard of living – in the Netherlands nearly 400 people to the square kilometre. There are countries with great numbers of people and a very low standard of living – India with over 150 people to the square kilometre and most of them impoverished. Yet, population growth, wherever it occurs, takes place within the one limited biosphere. All people must be provided for from resources, the limits of which are becoming increasingly well known. All must live in the same life-supporting layer of the planet, within which all other living things must also continue to exist.

We now know reasonably well the boundary conditions within which we must exist. The question of how much food might be produced from land and water under existing or potentially feasible technologies can now be roughly estimated. We know within reasonable limits the possible extent of available mineral and fuel resources. We know further that we cannot expand the area of land devoted to one use without reducing the area available for other purposes. We can calculate how many people could theoretically be supported on Earth. But we know that, in the upper limits of our estimates, the conditions for human life would not be pleasant if they were, indeed, tolerable at all. We also know that whatever limit we are considering it can be reached within a period of time that is distressingly short. We can clearly foretell the end of population growth on Earth. The question to be answered is whether we shall limit growth in time to have a world worth living in.

The control of technology

During the Second World War great numbers of people were made homeless and forced to live under conditions of severe privation. Great numbers of soldiers were in the field in many parts of the world experiencing and in turn creating conditions of unusual hardship and stress. Circumstances seemed right for the rapid spread of epidemic diseases (such as typhus, plague and malaria) which had plagued mankind again and again over past centuries – diseases borne by insects. At this time, however, new chemicals came into use, belonging to a group known as the organo-chlorines or chlorinated hydrocarbons. Best known of these, and the first to be used, was DDT. This initially had a remarkable ability to destroy insects. In tropical areas sprayed with DDT the malaria mosquitoes, formerly present in great clouds, were suppressed or did not emerge from their larval stages. The incidence of malaria was much less than the medical authorities had expected. The dusting of powdered DDT around the dwellings or on the clothing of troops or refugees killed the lice and fleas that could have carried typhus or the plague. The epidemics, which could have killed more than any battles, did not take place.

Following the Second World War, the use of DDT and other chemically related pesticides expanded enormously. Disease-carrying insects were reduced to low levels and the general health of human populations in areas treated with pesticides improved. Applied to agricultural fields, the new pesticides reduced the numbers of crop-destroying insects. Agricultural yields began to climb to new peaks. It appeared to some that the chemical industries had provided humanity with the means for destroying two of its oldest enemies – famine and plague. The new pesticides were regarded as an unmixed blessing to mankind.

But, all the while, other studies were going on. It was soon learned that the pesticides did not confine their action to the target insects – they poisoned all insects that were exposed to them, including the predators and parasites that had previously controlled the pests. Fish, birds, mammals, all forms of life that consumed the pesticides could be poisoned by them. It was found further that some of the pesticides such as DDT were virtually indestructible, they went on and on accumulating in soils or waters or animal tissues and their effectiveness as poisons continued. As time passed, the original pesticides used were found to have less and less effect upon insect pests – the insects developed immunity to them and this was passed on from one generation to the next. Other slower-breeding species, or species that occurred in lesser numbers, could not develop resistance so quickly. Birds that fed on animals or fish were among the first species to show serious effects. DDT affected their reproduction, they laid thin-shelled or shell-less eggs and their young did not survive. Many species were threatened with extinction. Fisheries catches were made useless in many inland and coastal waters when the fish were found to contain levels of DDT dangerous to human health. Some studies suggested that DDT was affecting the small floating plants, the phyto-plankton, of the oceans, on which all of the world's fisheries ultimately depend. In some places DDT, used in areas that had not been affected greatly by insect pests, produced great out-breaks of crop-destroying insects. The countries that had pioneered the use of DDT began to restrict or even to forbid its use entirely. What had once appeared to be a blessing now seemed to be a curse that could endanger man's future on Earth.

This example illustrates the problems of control over tech-nology. Chemical industries produce each year a great number of new products, each intended for some useful or beneficial purpose. However, when introduced into the biosphere among the great complex of animal and plant species on Earth, they often produce effects far different from those anticipated. Their well-intentioned use in one area causes serious problems else-where, or in some direction that had not been planned.

The automobile industries of the world produce motor cars which have been in high demand for personal transport. The petroleum industries produce the fuel on which they run. Motor cars have changed entire patterns of living for people in many parts of the world and have permitted major improvements in their daily lives. Yet, unplanned for was the shattering impact of great numbers of automobiles on the structure of cities, of the effects of roads and highways upon fragile and highly valuable lands. Anticipated least of all was the pollution of the air resulting from the inefficient combustion of fuels in the auto-mobile engine, which in some cities has serious effects upon human health, plant life and animal life. Originally the motor car was welcomed as a great boon conferred by industry to enrich people's lives. Now, in areas where automobiles are concentrated, their presence is deplored by city planners, architects, gardeners, farmers, foresters and all concerned with human health and the health of the biosphere. We are forced to devise or convert to less damaging forms of personal transport.

A great dam is built on a river to provide water adequate to bring a large new area of land into food production. The aim is the benefit of man and the enrichment of an economy. Engineers plan and construct the system of dam, reservoir and canals in such a way that the objective is achieved. Previously dry lands bloom with new crops. But failure to attend to lands upstream from the dam results in ever growing loads of silt being carried downstream to be deposited in the reservoir. The rate of silt accumulation threatens the life and value of the original construction. Downstream, the absence of a regular seasonal deposition of silt on the flood plain forces more expensive measures to maintain the fertility and structure of soils. The restriction of the river flow causes encroachment of the sea into agricultural lands of the delta, threatening the loss of farming lands perhaps equal in area to the newly irrigated lands brought into production by irrigation. The cutting off of the supply of chemical nutrients, formerly carried by the river waters, causes a decline in fishery production. Finally, in the newly irrigated lands, waterborne diseases spread rapidly and threaten the health of the population. One of the most dangerous of these is bilharzia, caused by a blood fluke which is harboured in its developmental stages by a freshwater snail. In some irrigated areas, one hundred per cent of the human population has been infected. Technology, applied without full concern for the environment, has brought, once again, environmental costs that threaten to outweigh the benefits conferred.

A pulp mill is built on the side of a river. It produces materials essential to the operation of modern society, employs large numbers of people and allows them to share in the economic benefits to be derived from this enterprise. However, the logs that are processed in the mill have been treated with a fungicide containing an organic mercury compound and this poisonous substance is released in the effluent. Initially, there is no problem, since the river waters dilute and disperse the materials released by the mill. But because the site is a favourable one, other mills are set up along the same river and also release their effluents into the waters. Fish began to die for reasons that are at first not known. Then people become ill from a strange disease that is finally diagnosed as a form of mercury poisoning. The mills are forced, at some increase in cost, to find some other means for arresting the decay of the water-soaked logs, and are forced to close while this is done. The problem is one of technology. It could have been anticipated and prevented, but it was not, in part because no one person or organization was responsible for estimating the total effluent load to be deposited by all mills in this area or for evaluating the potential dangers that could be involved in this process.

If the world population ceased to grow and technology were to continue to be employed as it has been in the past in an effort to raise the standards of living everywhere to a high level, there would be major and increasing problems of the human environment. Population control is regarded as one means of guaranteeing for the future an environment adequate to meet human needs and to provide the satisfactions for the human spirit that people everywhere would like to have. But population control, without control over technological processes, is obviously inadequate. It is now obvious that any important modification of the environment brought about by man's industry and activities must be evaluated in terms of its long-range effects upon human lives and upon the air, water and lands upon which human life depends. The narrow pursuit of limited objectives, tolerable when human numbers were few and human powers limited, has now become intolerable.

The control of land use

In many places in the world, human numbers are sparse and advanced technology does not exist. Yet environmental damage goes on at a pace that threatens the continued existence of people in those areas. Great areas of formerly productive land are converted into deserts. Areas that once supported valuable tracts of forest are cleared for agriculture only to have their soils eroded away or the fertility of the soil lost. For the short-term gain of a few years of farming, long-term damage is done which may be irreparable for generations to come. Great herds of wild animals are exterminated in areas where they had long existed in balance with the vegetation and soil. They are replaced by herds of livestock and for a time the people in the area feel that their lives have been improved. But then the plants are killed off by too many grazing mouths and too many trampling hooves staying for too long in one place. Soils are exposed to the force of wind and rain and wash or blow away. The domestic animals begin to die of hunger, the people feel increasing impoverishment and they move away to seek some other area that has been cleared, or perhaps they move toward a city where food and work is said to be available. There they find no place, but gather in growing numbers in sprawling shanty towns, without sanitation, without adequate water, without medical care, without enough food.

Elsewhere, in more advanced countries, cities sprawl without plan across the most valuable farm land, forcing the production of crops to less fertile sites where they are grown at greater expense for poorer quality. A government agency works at great cost to protect and manage a marshland for waterfowl production and human recreational use, while another agency expends its energies in plans to drain the same area. A park is preserved and set aside for human enjoyment at great expense and then the middle is cut out of it by a highway that destroys much of its original value.

The need for planning and control of land use, taking into account the capabilities of the land for sustained production and its capacity to support the kind of use intended, is widely recognized by governments, but they are often incredibly difficult to put into operation. Traditional ways of using land persist and peoples resist, sometimes with violence, any attempt to change them. Traditional attitudes toward age-old 'rights' to use land that is unoccupied force retreat and compromise on the part of authorities charged with protecting the land. In countries where planning and control over land use is to some degree accepted, planners with differing objectives often work in isolation from each other and all fail to consider the full range of existing or potential values that a land area represents.

If populations ceased to grow, and the adverse effects of technology were fully controlled, environmental damage would continue unless effective control over land use could be generally obtained. Such control must be balanced by an understanding of the biosphere and the ecological principles that describe its operation. A long-term concern for the total environment of man must take equal place in planning with short-term economic objectives.

The absence of information

The problems caused by growth of population, by lack of control over technology, and by lack of control over land use, all derive in part from our continued ignorance of the biosphere. Although science has made enormous and spectacular gains in some areas, and combined with technology permits us to place men and machines on the moon or to split the atom, in other areas, and indeed in most areas concerned with man's relationship with the biosphere, our knowledge lags. We are only beginning to understand the intricate workings of complex communities of plants and animals. The blunders we have made with certain pesticides or in the use – or misuse – of atomic energy have not been unmitigated, since they have permitted and encouraged the research that has given us more understanding than we should have had in the absence of these dangers. Yet, we do not really know what is required to protect our planet from unanticipated effects of our technological endeavours. And we do not know what effects our modification of environments is having upon ourselves.

It is particularly in the sciences that deal with living things – the social sciences concerned with man and his behaviour, and ecology, the science concerned with the complex nature of communities of plants and animals – that our knowledge is most inadequate. Furthermore, much knowledge that is acquired in one nation is not generally available to people in other nations. In addition, certain kinds of knowledge, such as measurements of the effects of human activities on the composition of the atmosphere or the oceans, are not the responsibility of one nation nor can they be carried out without the cooperation of all nations. The effects of pollutants, for example, wherever they are produced or released, must be studied on a global scale. Many of the poorer nations cannot afford to carry out the research and management needed for the proper care of their lands and resources, yet their activities ultimately will affect all nations. It is thus that the need for a major inter-governmental research and training programme directed toward the problems of man and the biosphere arises. Such a programme was called for by the Biosphere Conference and is now approved for operation by Unesco in cooperation with other international agencies. Without the information that this programme can derive, we may well continue to make the same or new blunders in our use of the environment and thus endanger the future of man.

Chapter 2
The Biosphere

The biosphere has been defined as that layer of soil, water and air upon the surface of the Earth within which all life exists and of which it forms a part. It is that combination of living things with their physical environment which is to be found from the depths of the oceans up to the lower limits of the atmosphere. Life and its environment cannot be separated. Each individual man, or animal, breathes air and, through breathing it, changes its composition; each takes in water in one form or another and gives off water containing waste products; each takes in foods built from soil minerals, water and air, and returns them in modified form to the soil, the water and the air. Equally, plants constantly take in air, water and soil minerals, and give off variously modified forms of these substances. An organism apart from its environment cannot exist. Even after death, the interactions continue in a different way. In the biosphere, therefore, a process of continual interaction goes on between life and non-living matter and energy. The nature of the air, water and soil in the biosphere is changed through their interaction with living things, and living things are in turn dependent upon the physical environment which they help to create and to maintain. It would be possible to have layers of atmosphere, rock and water on the Earth without life, but they would not make up a biosphere, and their chemical composition would differ from that which we see on Earth in the presence of life. The human race makes up only a small portion of the total living material in the biosphere. It is dependent upon the great variety of other animals and plants on Earth to maintain the environment within which human life is possible.

Structure of the biosphere
The atmospheric component

Extending for many hundreds of miles above the surface of the Earth, the atmosphere forms a layer of protection for all that takes place in the biosphere. It is the region of interaction between the molecules and atoms that form part of our terrestrial system and the energy and matter given off by the sun or reaching the Earth from outer space. In the atmosphere energy from solar radiation gives rise to enormous forces that power the Earth's weather systems, and through their cumulative effects

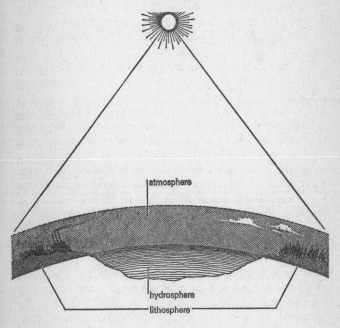

atmosphere

hydrosphere
lithosphere

Figure 2 The biosphere: the surface layer of air, water, rock and soil on our planet in which all life exists and of which all life forms a part

determine the Earth's climates. The enormity of this energy is to be seen on a small scale in the shattering force of a tropical hurricane. Much greater amounts of energy are involved, in a more diffused form, in the swirling masses of air that may be observed, through their cloud patterns, in the satellite photographs of the Earth taken from the higher reaches of the atmosphere – the cyclones and anti-cyclones that dominate the weather of the northern and southern hemispheres.

In addition to its role in distributing the energy of solar radiation over the Earth and thus determining the conditions for life on its surface, the atmosphere becomes directly involved in the chemical transactions of life. Atmospheric gases are incorporated in all living matter. Life, in turn, determines the composition of the atmosphere.

The Earth's weather, for the most part, takes place in the dense lower portion of the atmosphere, known as the troposphere, extending up to a height of no more than sixteen kilometres. Above that extends a region of cool, sparse air that was not long ago considered to be the 'top' of the atmosphere – the stratosphere. But the space age has extended our knowledge outward. Beyond the stratosphere are other layers of thinly dispersed gases, the mesosphere, thermosphere and finally exosphere, which blend into the relative emptiness of interplanetary space. All of these help to shield the Earth from life-destroying radiation – ultraviolet rays from the sun, and cosmic rays from the outer reaches of space among its various forms – that would otherwise blast its surface.

The composition of the Earth's atmosphere by volume amounts to a calculated 78 per cent nitrogen gas, plus nearly 21 per cent oxygen, with the balance divided mostly between the inert gas argon and carbon dioxide. A very small fraction is composed of other gases which include hydrogen, ozone, methane, radon, helium, neon, krypton, xenon, sulphur dioxide, hydrogen sulphide and ammonia. Most of these play no obvious role in life processes, but some are extremely important to life. For example, ozone, an active form of oxygen (three atoms instead of the usual two in each molecule), is damaging to living things when in direct contact. However, it forms an important shielding layer in the stratosphere that absorbs ultraviolet light and other dangerous radiation. In addition to these gases the atmosphere contains water vapour, which varies in amount from place to place depending upon the evaporation from the water surfaces of the globe and on the precipitation in the form of rain, snow, hail and sleet. The atmosphere also contains suspended solid particles, particulates or aerosols, dust, smoke and various salts. Volcanic action in particular, but also the effects of man's use of the Earth's surface, can cause great variations in the amount of particulate matter in the air as well as in the amounts of carbon dioxide, sulphur dioxide and other gases.

The composition of the atmosphere, however, reflects the interactions that take place in the biosphere. The atmosphere has been modified by life into its present form. Some idea of what an atmosphere without life would be like is revealed by recent explorations of the planet Venus. Nearly 95 per cent of the Venusian atmosphere is carbon dioxide, approximately 4 per cent is nitrogen, whereas the balance includes water vapour, oxygen and other gases. Temperatures on the surface of Venus are hundreds of degrees higher than on Earth, because the dense layer of carbon dioxide traps the longer wave lengths of solar radiation, heat rays and infrared rays, much as the glass of a greenhouse traps the heat from sunlight and prevents it from being radiated back into the air. However, if the total amount of carbon dioxide on Venus and on Earth are calculated, they are approximately the same. The difference lies in most of the Earth's carbon dioxide being combined in the rocks and liquids of the Earth's surface – in the carbonates of sedimentary rocks and in the deposits of coal, oil and natural gas included within these rocks. Life processes, especially photosynthesis, carried out by green plants over the thousands of millions of years that life has been present on Earth, removed carbon dioxide from the atmosphere of Earth and, in turn, added oxygen to that atmosphere making possible the existence of all animal life, including man.

On Venus there is no liquid water, because of the high temperatures, and there is relatively little water vapour. On Earth, of course, quite the reverse is true. Part of the reason lies in the existence in the Earth's atmosphere of a cool stratosphere in which temperatures fall to about −60 °C. Such cool air can hold little water vapour and acts as a barrier to the upward movement of water, which is therefore largely confined to the troposphere and the surface. On Venus, for reasons that are still obscure, the stratosphere is much warmer. Water vapour can move upward through it to higher altitudes where it becomes bombarded by ultraviolet and other short-wave radiation. This breaks it down into its component parts, oxygen and hydrogen, of which the hydrogen, because of its light weight and the high velocities it builds up in the elevated temperatures of the thermosphere, can escape into outer space.

Thus the composition and structure of the Earth's atmosphere shields and favours life, which in turn has modified the atmosphere to make it increasingly favourable to the continued existence of life. Or so it was, until recent interventions by technological civilization which will be discussed later.

The hydrosphere

If all of the water in the biosphere were to be concentrated in liquid form and distributed evenly over the surface of the Earth, it would cover the entire Earth to a depth of 3·2 kilometres. This hydrosphere would then be a continuous layer resting below the atmosphere and could form an interesting environment for all marine forms of life. It would of course provide no home for those creatures that have adapted to life in the open air or in fresh water. Fortunately the hydrosphere has been disrupted by the emergence of the Earth's continents and islands which now occupy over 30 per cent of the planet's surface and have at various times in the past occupied slightly more or sometimes much less. Here all of the terrestrial forms of life have evolved, as well as those aquatic forms that have learned to live apart from the salty waters of the world's oceans. However, life, wherever it exists, remains as dependent upon water as it was in the earlier ages when it existed only in the oceans. The movement of water from oceans to the continents by way of evaporation into the atmosphere and then precipitation on land, leading back to the oceans by way of streams, rivers and underground channels, is known as the hydrologic cycle. On its proper functioning depends the continued existence of land and freshwater life. A major inter-governmental programme, the

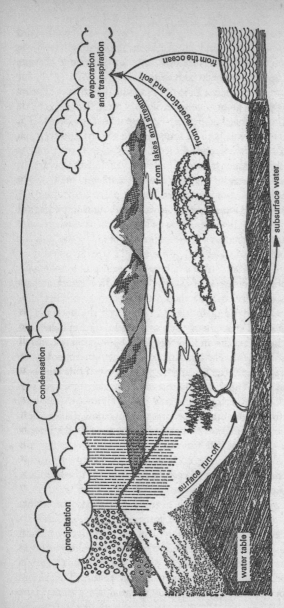

Figure 3 The hydrologic cycle (water cycle): water moves from the ocean to the atmosphere to the land and then returns once more to the ocean

International Hydrological Decade, is now devoted to studying the operation of the hydrologic cycle over the entire globe. Many results have already been produced and our knowledge of water and its movement has been increased.

There are an estimated 1360 million cubic kilometres of water on Earth. Of this, at any one time, slightly more than 97 per cent (1320 million cubic kilometres) is present in the oceans. The next largest reservoir is to be found in the glaciers and ice-caps, mainly in Antarctica and Greenland. These, containing 30 million cubic kilometres, could raise the level of the oceans by a few hundred metres if they were to be completely melted. The next largest amount of water, less than 1 per cent of the total, is in underground storage, ground water, amounting perhaps to 9 million cubic kilometres. Freshwater lakes, salt-water lakes and inland seas such as the Caspian, contain some 230 000 cubic kilometres of water. All of the world's streambeds contain only 1250 cubic kilometres, whereas the atmosphere holds only 13 000 cubic kilometres of water. These calculations are approximate, but serve to show the relative balance.

It is not the static quantities of water, but the dynamics of water movement that are of such overwhelming importance to life. Both in the oceans and beyond their boundaries water is in constant movement. Nearly 450 000 cubic kilometres of water are evaporated from the ocean each year. Most of this falls back into the oceans once more in the form of rain or other precipitation. However, over 100 000 cubic kilometres of water fall upon the continents in the form of rain, snow, hail or sleet. The greater part of this is evaporated or transpired back into the atmosphere, some to fall again in precipitation. From the continents, each year, nearly 40 000 cubic kilometres run off through streams and rivers. These carry quantities of dissolved and suspended materials which enrich the waters.

There is no shortage of water, in any absolute sense, for human purposes, either now or in the foreseeable future. Similarly, there is adequate water to support all of the primitive abundance of life that Earth contained before man's technology evolved. There are, however, shortages of useful water – not too saline, not too polluted – in many parts of the planet. The natural distribution of fresh water is highly unequal, since the currents of the Earth's atmosphere bring excessive amounts – over 10 metres per year in precipitation to some parts of the globe such as tropical mountains in the trade-wind belt – and virtually no rain at all to other areas – the Atacama desert of Chile, parts of the Sahara, or the Namib desert of south-western Africa.

It is particularly the water that falls from the atmosphere to the ground that becomes useful to terrestrial life. This enters the soil, from there it can become involved in the life processes of land plants, and through them, animals. Or it runs off from the soil to form streams, rivers and lakes – the aquatic habitats in which a rich variety of plants and animals find a place to live. It is also the water that reaches the ground that proves directly useful to man. He draws on it to grow agricultural crops, and this is the most demanding of uses since water is either incorporated in plant material, transpired through the leaves of the plants, or evaporated from agricultural soils. The water that runs off the soil surface may be impounded by dams, in reservoirs, from which it is used to provide electric power or to irrigate crops grown on dry lands. Some of it is used for the water supply of urban and industrial areas, in which it becomes involved in a wide variety of processes and, if properly managed, may be re-used many times. Through all of these processes water may become polluted – laden with an excess of inorganic chemical waste products or with the organic wastes of human sewage, food-processing plants, abattoirs, etc.

All living things are composed mostly of water and all require either immersion in water or the provision of large quantities of water, either directly or combined in food, if they are to keep functioning. It is estimated that it takes ten tonnes of water to produce one tonne of animal tissue. One way or another such quantities of water must be provided if human and animal life is to exist. But some ways of providing for such needs are more expensive in their water requirements than are others. Thus it takes a thousand tonnes of water, evaporated, transpired, or used directly, to produce one tonne of sugar or maize from irrigated lands. Where water is in short supply irrigation agriculture is an extremely wasteful process and can only be justified by the high value of the crops produced or by serious human needs in areas dependent on this form of land use.

The lithosphere

The rocks of the Earth's crust and the soils derived from them form the lithosphere – the foundation from which all living things ultimately are constructed. With the exception of hydrogen, oxygen, nitrogen and carbon, which may be derived from air or water, all of the other chemical components of life originate in the crustal rocks of the Earth. Yet these rocks, in turn, are formed, or are changed and modified, by the action of living creatures.

The core of the Earth differs markedly from the Earth's surface. It is molten because of the intense heat, yet extremely dense because of the high pressure, and is composed primarily of the two metals, iron and nickel. By contrast, the Earth's crust is solid and contains a much greater variety of chemical elements. The deeper part of the crust that underlies the continents and the oceans is primarily composed of basalt – an igneous (fire-formed) rock that results from the melting and subsequent cooling and crystallization of minerals beneath the surface of the ground. By contrast the continents are of lighter rock, predominantly granites, which float and move about on the denser basaltic floor of the Earth's crust. It is now generally accepted that the continents have drifted apart from a time hundreds of millions of years ago when they were connected together. The obvious fit of the angle of Brazil into Africa at the Gulf of Guinea is no accidental coincidence in appearance.

The surface continental rocks are of the greatest interest since these for the most part are those that interact with the biosphere. Exposed to weathering from sunlight and temperature change, the action of water and of wind, they gradually break down and decompose, forming a substrate that can be modified by plants and animals to form soil. Soil is a place of interaction between the atmosphere, the water from the hydrosphere, the lithosphere and the living organisms that these all support and by which they are in turn affected and changed. Soil, when leached by rain or eroded away by wind or water runoff, adds minerals which support life in fresh waters and, accumulating over millions of years, maintain and support the richness of the oceans.

Apart from the soils and surface rocks, the lithosphere was once considered as something separate from life – inorganic and sterile, existing now as it existed before life began on Earth. But now the sciences of biology and geology are converging as they explore the origins and effects of life upon the surface rocks of the planet. They see the role of living things in modifying and changing the surface rock, adding their components to the accumulations of sediments washed or blown to low-lying areas, thus contributing to the formation in time of sedimentary rocks. These, in turn, sink deeper in the crust, are heated and pressurized into the other forms of crustal rocks – metamorphic and igneous. Eventually old sedimentary materials, the 'products of past biospheres' emerge once more on the surface in the form of 'new' igneous rocks – granites, basalts and their relatives. In the words of the Soviet scientist V. Kovda and his associates, reporting to the Biosphere Conference, 'The atoms of almost all the chemical elements (on the surface of the Earth) have passed through living matter innumerable times in the course of complex cycles. The appearance of the planet has greatly changed and it may be considered that it is living matter that has determined the composition of the atmosphere, sedimentary rocks, soil and, to a great extent, the hydrosphere.'

The soil is the vital component for all terrestrial life. It is built from rock by the action of weather and life. It is formed, shaped and enriched by living things – bacteria that take nitrogen from the air and add it to the soil in the form of nitrates are an example. Here plant roots take hold and draw nutrients from the soil, but in turn add various elements to its composition. The action of wind and water modifies soil by erosion or drift or the leaching of minerals from the surface by the downward movement of water. A layer of life, of plants and animals, arrests the action of erosive forces and restores minerals once more to the upper soil. Destroy the living cover of the soil, and you risk its destruction or loss. If soil is lost the productive capacity of the terrestrial part of the biosphere is largely lost also. It is on this capacity that human life still depends.

The vital component

Some three thousand million years ago life first appeared on Earth. Perhaps in the rich 'organic soup' of some shallow sea, that mixture of nitrogen, carbon, hydrogen and oxygen, along with some smaller amounts of associated and essential atoms, was joined together in a combination unique among all the previously existing forms. The new entities could grow by taking in and recombining the organic molecules that surrounded them, that is, they could feed upon the non-living world, and perhaps upon each other. Most important, they could divide and reproduce, thus perpetuating themselves.

Over time, these primitive forms of life would be bombarded by the radiation present in their environment. At times this could be fatal, but in other instances it may have given rise to slight shifts in the atomic or molecular structure of these new organisms, bringing a change or mutation which permitted the evolution of a new form of creature adapted to a slightly different environment and way of life. Cumulative changes of this sort could lead to the eventual appearance of organisms markedly different from their ancestors. Certainly the most important of these changes were those that enabled their possessors to develop the process known as photosynthesis. In this process the gas, carbon dioxide, present in abundance in the primitive atmosphere, and the liquid, water, equally or more abundant, are combined to form a simple organic molecule, the sugar, glucose. To accomplish this requires energy, but in photosynthesis this is obtained from the sunlight that pours down upon the lighted layers of the ocean and is captured by the chlorophyll that occurs in the surface cells of green plants. The first organisms able to carry out photosynthesis would be freed from dependence on either the preformed organic materials that had been present on Earth before life appeared, or upon the necessity to feed on other living things. These organisms, the first green plants, brought hope that life could continue as long as Earth and sun endured. They formed the basis on which all future life was to build since they alone could take the relatively simple elements present on Earth and form them into complex living structures. They alone could capture and store the energy of sunlight and make it available for the support, not only of themselves, but of all things that fed upon them.

In the early oceans this innovatory change gave rise to the green plant plankton, the floating 'grass of the sea' on which all marine life depends. Life could then further proliferate and change. Hundreds of thousands of species were to evolve and spread into all available habitats from the ocean deeps to the edges of the land. Before the first land creature appeared on Earth, most of the forms still to be found in the oceans had already appeared, at least in a primitive form.

Perhaps five hundred million years ago life first moved from the oceans. Some creatures acquired gradually the ability to resist the desiccation that threatens creatures exposed to the atmosphere, carrying their own aquatic environment with them internally. Others, moving up estuaries, learned to tolerate gradually decreasing salinities and to occupy the freshwater streams, swamps and lakes. In these new environments a further 'explosion' of evolution took place as species developed which could exploit and make use of the wide range of terrestrial and aquatic environments. It is estimated that at least three million species of plants, animals and micro-organisms have appeared on Earth, an enormous richness of life, able to change and modify the planet. One of these species is man.

The arrangements of living things

Each species of animal or plant is characterized by a unique ability to make use of a particular kind of environment, to feed upon certain things and in turn to be fed upon. Each species, therefore, is said to have its own ecological niche on Earth, which no other species occupies. The ecological niche defines the place in the environment for a species, and its relations with other species.

Species which have similar tolerances for climate, soil and other characteristics of the physical environment, such as the amount of water, the topography, etc., tend to come together in a common habitat, within which each occupies a separate niche through which it interacts in various degrees with the others. Such a group of species forms a community, or a *biotic community*. A woodland grove with its associated animal life, a meadow with its animals, a forested mountain slope, the plants and animals that live in a temporary pond, a marsh or a lake, each of these represents a biotic community.

A biotic community always contains a certain array of organisms. There must normally be green plants, for these provide the food, or energy, for the entire community – obtaining it through photosynthesis, which combines carbon dioxide, water and sunlight energy – enriching it in the form of more complex molecules built up through the addition of minerals extracted from the soil or water by the plant. Next, there are the animals that feed upon plants, the herbivores. These include micro-organisms, small animals of various kinds, a wide variety of insects, and usually some birds, mammals and other vertebrates such as reptiles, fish or amphibians. Next, are the animals that feed on animals, the carnivores, including insects, fish, birds and mammals. Finally, there are the creatures that feed upon the dead remains of plants and animals and break these down into chemicals that are added to the soil or water. Mostly these are small bacteria or fungi, but they include larger animals as well. The visible components of a biotic community are most impressive, but the minute life of the soil is present in great abundance and variety and, in rich soils, may be more totally productive than the large and obvious living things. It is not uncommon to find a tonne of bacteria in the topsoil of one hectare, and these are among the smallest of soil organisms.

Biotic communities vary in size and complexity from the simple communities to be found growing on rocks, made up of a few kinds of lichens or mosses with some associated microscopic forms of life, or the more complex form that may be observed under the microscope in a drop of pond water, all the way up to the enormous complexity of the communities found on coral reefs in tropical oceans or found in rain forests in tropical lands.

The larger communities of the Earth that occupy major regions and share a common tolerance to a regional climate are known as *biomes*. Examples of these are the desert communities of the Earth, the grasslands, the tropical rainforest, the temperate zone broadleaved forests, the coniferous forests of the cooler, northern regions, or the tundra of the arctic. Biomes have common characteristics that make them similarly responsive to various forms of management by man. A tropical rainforest biome in Africa or America will have similar characteristics, problems, and potentialities for human use. Nevertheless, within any biome, different complexes of living things will be found and each will have its own unique potentials and present similar problems when people attempt to make use of it.

Subdivisions of the biosphere

Each different community on Earth, whether major or minor, forms the basis for recognizing the various subdivisions of the biosphere that are known as *ecosystems*. An ecosystem, in simple terms, is the combination of a biotic community and its physical environment. We have already noted that it is impossible to separate living things from their non-living environment. A meadow cannot be separated from the soils and water that support it, the animals, large or microscopic in size, that live in it, or the atmosphere that surrounds it – in combination they form a meadow ecosystem.

Each ecosystem is a functioning, dynamic system in which the lithosphere, atmosphere and hydrosphere interact with life. Each is an open system in which the source of energy, sunlight, is external and in which the source of water, the atmosphere in most terrestrial instances, but in some cases stream flow or ground water flow, is also external. Even in the oceans, the movements of ocean currents, or the vertical movement of water in ocean basins, brings water into and out of any ecosystem. The principal source of minerals is usually internal in terrestrial ecosystems, although some are added by air and water movement, but may be external in marine or freshwater systems. The ecosystem itself has an internal integrity – energy and matter move within it in orderly and more or less predictable patterns. Chemical materials are commonly recirculated within the system and re-used again and again.

Ecosystems may be natural or man-made. In the former, the various components are arranged largely without human intervention. In the latter, plants and animals may be domestic or wild but they are arranged in patterns considered more useful or more pleasing to man. Such domesticated ecosystems range

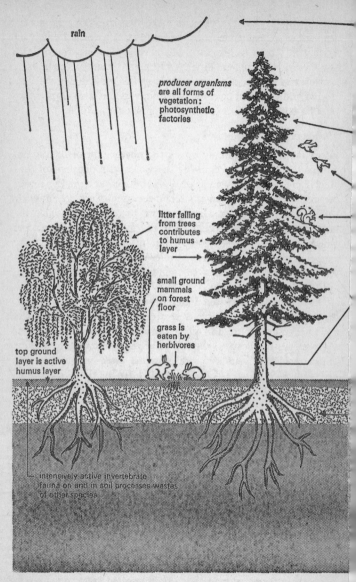

Figure 4 An ecosystem and its functioning

Text labels within the figure:

rain

producer organisms are all forms of vegetation: photosynthetic factories

litter falling from trees contributes to humus layer

small ground mammals on forest floor

grass is eaten by herbivores

top ground layer is active humus layer

intensively active invertebrate fauna on and in soil processes wastes of other species

active influence on flora
and fauna of seasonal variations
in temperature, light
and moisture

sunlight

consumer organisms
are all forms of animal
life; they process
organic matter

foliage serves as cover
and as nesting sites
for birds, e.g. owls,
which eat small rodents

aspen
or
birch

vegetative shoots
are eaten by squirrels
and some birds;
squirrels also eat cones;
crossbills eat seeds
from cones

leaf-eating insect
larvae drop calcium-rich
juices on field layer,
their population
is regulated
by insect-eating birds

lichen on bark is a
micro-pioneer ecosystem

urine and droppings
from grazing deer
fertilize field layer

deciduous trees
improve acidity of soil

mycorrhizal fungi
in symbiosis with roots
of pine trees

bedrock slowly being dissolved
by root tips of trees; trees
themselves are pumps of
nutrients from below ground

from city parks or gardens to the agro-ecosystem of farming land or tree plantations used for wood production.

Functioning of the biosphere and its ecosystem

When there was no life on Earth, no biosphere, the great outpouring of energy from the sun had only a scant effect upon this planet. For a brief time, while exposed to the sun, the surface of the Earth would be heated. But this energy would quickly be lost, as it is now from the moon, re-radiated back into space after having accomplished only a transitory effect. The Earth would differ from the moon in that its atmosphere and oceans would be heated, and these would hold the solar energy for a longer time before it was radiated back into space, but the results would be the same. In the absence of life, therefore, the solar system could be regarded as a system through which energy was increasingly dispersed from its central concentrated source to be lost in the immensities of space. Entropy, that dispersion of matter and energy to an increasingly scattered and randomized state, would increase.

On the Earth, however small it may be in the vast space of the universe, this process is arrested. Solar energy is halted in its movement outward into space and concentrated by living things to be stored for future use. Randomization and entropy are decreased by life. With the increasing complexity and organization of living communities there is a growing capacity to capture, concentrate, utilize or store energy that would otherwise be scattered and wasted. This is one of the greater miracles of life. Its principal agents are the green plants – from the humble algal cell that forms a minute part of the mass of floating plankton in the oceans to the giant redwood tree or the enormous kelp plants of shallow seas. Each, while it lives, captures energy from the sun, converts it into bonds that hold its molecules together, and stores it for future use. Each uses some of the energy it captures, but stores a surplus, which can then support other plants, animals or, ultimately, the micro-organisms of decay. Dead plant material in the humus of the soil, the trunks of trees or the organic debris accumulating at the bottoms of lakes or oceans, represents energy captured in the past. When we burn wood we are releasing solar energy stored during perhaps several centuries of plant growth. In the organic remains in sedimentary rocks, energy stored long in the past may be found. The deposits of coal, oil, oil shale, natural gas and other fossil fuels represent energy captured from the sun by green plants in early eras of geological time, some of it dating from the days when life was young. The biosphere as a whole is an enormous storehouse of energy. So long as the life within it continues to function in all its diversity these stores are renewed. Unfortunately, technological civilization has placed unprecedented pressures upon these natural storehouses of energy, draining them off at rates that far exceed the biosphere's capacity for restoration.

Energy flow

Within each ecosystem and community, energy is passed from one organism to another along pathways that are known as *food chains*. At the base of each chain is soil, sunlight, water and air. Utilizing these and transforming them into useful stores of energy and materials are the green plants – the producer organisms of the system. Feeding on the producers are the consumers – all the animals and the non-green plants. Within the consumers are further subdivisions: herbivores that feed exclusively on plants, carnivores that feed on herbivores, and sometimes carnivores that feed on other carnivores. Living off the producers and consumers alike are the reducer organisms, the creatures of breakdown and decay. Each depends upon sunlight energy stored by plants, each uses and stores some of this energy, which then becomes available for other users. But in each transaction energy is lost. The most efficient herbivore cannot store more than 20 to 30 per cent of the energy contained within green plants – the rest is lost in waste products, or in the heat generated in its consumption, digestion and metabolism. Similarly carnivores are little more effective and store within their bodies only a portion of that energy available in the meat that they eat. Within any ecosystem, therefore, a dynamic flow of energy goes on at all times – some carries out its work and then is lost to the system, some is stored for future use, some locked away in long-term storage.

Since man is dependent upon the food energy stored in natural ecosystems, or those artificial domesticated systems that

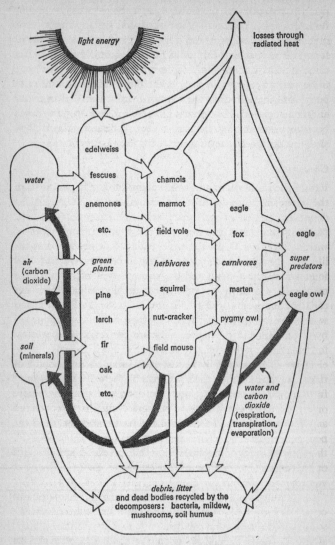

Figure 5 Food chains and an ecological pyramid in the Alps

he puts together in his fields and gardens, his supplies depend in part on the position he occupies in a food chain. When he functions as a herbivore, eating plant foods, he has more calories of energy available for his personal maintenance. When he functions as a carnivore, eating meat, he obtains fewer of the calories poured down originally in sunlight, but he may obtain a more palatable and nutritious diet. Where great quantities of people are supported on few resources, as in densely populated and little-developed countries, a major reliance on plant foods becomes essential. Little meat can be made available because sheer quantities of calories are essential for mere survival.

Chemical cycles

Equally important to the flow of energy through food chains is the flow of chemical materials, the nutrients on which life depends. Unlike energy, these tend to be used again and again, passing from a food chain back to the soil, water, or air, from which they are once again captured for use in another chain of life. As already noted, nearly every atom or molecule to be found in a human body today has been used before by other men at other times or by animal and plant life. When we breathe the toxic fumes released by motor cars or heating plants in the world cities today, we are inhaling molecules strung together by living creatures in the swamp forests of the Carboniferous or formed by the plankton in some ancient sea. This recycling of elements is a function of ecosystems upon which their continuation depends. A key role is played by the reducer organisms: from the larger scavengers that tear apart the remains of dead organisms to the small bacteria that make the finer subdivisions and return chemicals to the soil for future use. Most of the pollution problems of mankind today result from the choking of these natural recycling processes with excess amounts of waste or the poisoning with toxic compounds of the organisms involved in recycling.

In recent years the capacity of living organisms to concentrate certain elements to a degree far in excess of that occurring in the soils or water from which they are obtained has attracted much attention. Research on the fate of radio-iodine, released by nuclear explosions or inadvertently in other uses of radioactive materials, revealed the high concentrations that were reached in the thyroids of animals high in the food chain compared to that to be found in plants or in the soil or water. Generally, the higher a position occupied in a food chain, the greater opportunity for the concentration of chemical materials. Thus the effects of the chemical pesticide, D D T, which is both toxic and persistent in the environment, were noted in meat-eating and fish-eating birds before they could be observed in species lower in food chains. Although D D T may be in low concentration in the ocean, it becomes accumulated in a greater concentration in plankton. In turn the fish that feed on plankton concentrate it further, and those that feed on these smaller fish concentrate it still further. A pelican or cormorant, feeding on larger fish, receives a dangerous concentration of the pesticide with the result that successful breeding is brought to a halt.

Growth and change in ecosystems

Living systems are never static but always in a dynamic state. At times a so-called 'steady state' may be reached in an ecosystem, in which the inflow of energy and materials is balanced by the outflow, and a relatively stable condition is maintained. At other times, however, an ecosystem may be in a state of active growth and accumulation and the input exceeds any loss from the system. Still other systems may be in a state of decline, with loss exceeding gain – excessive cropping or mining of ecosystems by man can bring this about and destroy a healthy, functioning biotic community, creating a barren waste in place of a thriving ecosystem.

Biotic succession

The process of growth and change known as biotic succession has been particularly well studied in the ecosystems of the land. Any area of land that has not previously supported terrestrial life – a lava flow, a talus slide formed at the base of a cliff, an exposed river bank or lake bed, will be invaded initially by relatively few species of plants and animals, those most hardy and adaptable. A 'pioneer' community will develop which will modify the habitat, tame the ground and make it more suitable for less hardy and adaptable forms to live. These will in turn replace the pioneers and form a different 'middle successional' community which in its turn will change and modify the environment. In the classic process this will give way in time to a relatively stable end product or 'climax' community, which will then tend to hold the ground, maintaining itself in a more or less steady-state condition over long periods of time.

When communities are modified by man, as when a forest is logged, a field is cleared, or a grassland grazed by domestic animals, the processes of succession are reversed and the ecosystem is set back to an early stage of growth. If it is then allowed to recover, it will again move through pioneer and middle successional stages and on toward another climax. This capacity of an ecosystem to recover from disturbance and replace itself through successional processes forms the basis for the principle of sustained yield and allows for rational use of biotic communities by man. So long as the disturbance is not too great and the basic environment is not disrupted, the damage heals over, new crops of trees, forage or wildlife are produced and they in turn can be harvested. But always the soil must be maintained, water must continue to be available, and a breeding stock of animals, or a seed stock of plants left to re-populate the area.

Many communities that man seeks to perpetuate for his own pleasure or use are those that occur only as successional stages in nature. Complete protection from disturbance in such communities leads only to their disappearance and replacement by climax forms. To maintain them the disturbing process that keeps the climax from taking over must be kept up. A notable example is to be found in pine forests. Most of these will be replaced in time by hardwood or mixed forests unless they are periodically burned over by lightly running ground fires or kept clear by some other process. With complete protection from disturbance, they will disappear.

Population growth

In addition to the overall processes of growth and change that affect total communities and ecosystems, each population of animals and plants within the community grows and changes – the change in the community is the sum of all of those changes within species populations. Each species, when it manages to locate itself in a favourable and previously unoccupied habitat, tends to increase in numbers. At first the rate of increase is most rapid, but because initial numbers are small, the total numerical increase is small. Two mice produce a litter and become eight mice. As numbers accumulate the percentage of growth in the population tends to decline, but the overall numerical increase becomes greater. Eight mice pair off and produce litters totalling twenty-four, bringing the population to thirty-two. The sixteen pairs produce ninety-six new mice and the population grows to one hundred and twelve. The percentage increase is less but the total increase is much greater than initially.

So long as the space, food supply and other requirements for the species is large in relation to the number of individuals, populations tend to increase at a rate approaching what is called their *biotic potential*, the maximum number of new individuals that can be produced by a species population in a given unit of time. However, as space or other necessities begin to become limited, some individuals in the population will receive less than they require to maintain maximum health and efficiency. The breeding of these individuals will decline or they may fail to breed at all. Loss of individuals through mortality will begin to increase. When this occurs, the rate of growth falls off and the population is said to be encountering *environmental resistance*.

Eventually, because no habitat can be unlimited in space, food or other needs, the entire population begins to encounter the limits of growth. Well-situated individuals will still remain healthy and live out normal life spans while producing young. Most individuals will be less well off, they will die young, or be unable to produce many young. The total gain in the population from the birth of young will be balanced by the loss from the mortality of other animals. Population growth will therefore cease, the population will have reached the limits of the environment, known as the *carrying capacity* for that species.

If the growth of such a population is plotted on graph paper with the number of individuals on one axis and time on the other, the curve described will tend to be S-shaped (or sigmoid). This is the logistic curve of population growth and reflects the initial, low numerical increase, the subsequent steep climb in numbers while the population is increasing at almost its biotic potential rate and the final levelling off which occurs when environmental resistance begins to equal the biotic potential – when birth rates decline and mortality increases.

Any ecosystem has a carrying capacity for any species population. No population tends to grow indefinitely, each will be levelled off by shortages of space, of food, of adequate protection from weather, of water or of other requirements. Some species tend to level off in numbers before the limits of their environment are too closely approached – they are self-regulating and individuals do not tolerate excessive crowding. Many of the larger carnivores are of this category – wolf packs or jaguars maintain their separate hunting territories. Some species are limited in their numbers by the effects of other species that prey upon them. This is the basis for biological control of certain insect pests. Still others that lack effective predators and have not acquired or have lost self-regulating behaviour must be controlled by other forces in their environment, by malnutrition and starvation, by disease, by various catastrophes. Deer herds freed from predators build up to high levels and die from malnutrition. Lemming populations in the Arctic build up to excessive levels and then die back to few survivors. Unfortunately, many segments of the human species appear to fall into this latter category – the most painful to the individuals and the most destructive to their habitat.

Biomass and productivity

The total weight of living material within any ecosystem is known as the *biomass* of that ecosystem. This may be broken down into the biomass of plants, of animals or of species and groups within that category. The biomass is a measure of the past growth of life within that ecosystem. When an ecosystem matures and reaches a steady state, the biomass tends to remain relatively constant – thus in an old forest the total biomass of vegetation will not change much from year to year, growth will be balanced by death. In a young forest, however, biomass will increase rapidly from year to year, with the rate of increase declining as maturity is reached.

Biomass varies greatly from one kind of ecosystem to another, depending on whether the requirements for living things are present in abundance or are restricted. Good soil, warm temperatures, abundant moisture, favour plant growth on land and, where these are all present, biomass will be great. Increasing aridity, excessive cold or the absence of a supply of chemical nutrients in the soil will result in a lower biomass being supported. In terrestrial communities the higher biomasses will be found in the humid, lowland tropics. However, as one travels in the tropics from humid to arid regions, water will become limiting and biomass will decrease, reaching an extremely low level in tropical deserts. Similarly, as one ascends tropical mountains, the increasing cold, which affects also the availability of moisture, will cause a decrease in biomass, and the higher mountain peaks will be almost barren of life. Travelling from the tropics northward or southward toward the poles, one will also find a decrease in biomass, reaching very low levels in polar regions.

Biomass represents the reserves of chemical nutrients and energy stored within living material. The human race, initially, gained the materials it needed for existence by tapping these living storehouses and using them for human purposes. But biomass, as such, is far less important for man's objectives than its counterpart – *productivity*. Productivity is the rate at which an ecosystem produces new organic material and through this stores energy. This also varies enormously between ecosystems. The natural ecosystems of the humid tropics, if compared with those of any other major biotic region on Earth, have both higher biomasses and higher productivity. Arctic and desert ecosystems, by comparison, have, on the average, low biomasses and low productivity. In the oceans, warm shallow tropical seas that support coral reef communities have high biomasses and high rates of productivity. Warm estuarine regions, where rivers enter the sea through a series of swamps or marshes, are notably high in productivity although their biomasses (or standing crops) may not be excessively high. Open oceans are both low in biomass and in productivity, since the chemical elements needed to support life are in low supply.

It is important not to confuse productivity in the total biological sense with agricultural productivity or yield. Humid tropical ecosystems are often poor producers of agricultural crops once the natural forests have been removed. It is essential, also, not to confuse biomass and productivity. The former, when high, is large and conspicuous, the latter may not be. The invisible algae in a pond commonly produce more new plant material (which is then used almost immediately as food by other organisms in the pond), than do the large and conspicuous rooted plants on the bottom or edges of the pond. An elephant has a large biomass, but produces relatively few young, which then require many years to grow and mature. The duiker, a small antelope with fast growth and high fertility, has a low biomass and may be far more productive. Given equal opportunities, the rabbits in an area will convert more plant material into animal protein than will the elk population. Invisible soil organisms will often produce far more new organic material per year than the conspicuous shrubs or trees that tower above them.

The limits of human exploitation of any community or ecosystem, be it wild or domestic, are determined by its productivity. Maintaining an ecosystem in a high state of productivity is one of the greatest challenges to man and of most importance to his long-term welfare. Activities which degrade productivity – soil erosion or loss of soil fertility for example – lower the carrying capacity of any area for man as well as for other forms of life.

Diversity and simplicity

Any species uses an environment in its own way. None are as versatile as man and even human abilities to use an environment are limited. Pine trees growing in a cleared area of ground will make use of some of the resources of soil, air and water – but if only pine trees grow there and nothing else, many of the resources will go to waste. Cows grazing in a field will eat certain plants, or portions of plants, and neglect others. If sheep and goats are added, a much wider range of vegetation will be utilized. The twenty or more species of wild grazing and browsing mammals that sometimes occur naturally in African savannas, from giraffe that feed on the crowns of trees to elephants that feed on their roots, make far more efficient use of the vegetation resources than would an equal number of animals made up of only several domesticated species. As a generality, therefore, it can be stated that the greater the diversity of species in an ecosystem, the more efficient will be the use of the physical resources of that system and the higher the overall productivity.

Over the millions of years during which animal and plant evolution has taken place, species evolved to fill the various ecological niches that were available in the environment. Each ecosystem developed an increasing complexity and diversity as its resources were exploited by more and more species. Nevertheless, the number of species able to adapt to various types of environment varied with the factors present in those environments. Few species can tolerate the aridity of deserts or the extreme cold of high mountains and the polar lands. Species diversity is therefore low in such regions, although the number of individuals of any one species may be very high. By contrast, thousands of species will occur in a small area of humid tropical forest. Species diversity tends to reach a maximum in tropical rain forests on land and tropical coral reefs in the oceans.

It is an essential characteristic of ecosystems that the species within them interact with one another. Some are parasitic, living on and at the expense of another species, such as fleas on a dog. Some have mutualistic relations – they benefit one another, such as tick birds and African buffalo. Some species will depend on another without either harming or benefitting it, as do the orchids that grow perched on the branches or trunks of rain forest trees. However, it is the direct food relationships between species, the predator–prey relationships, that play a major role in determining levels and balances between species populations. If there are few species in a community these balances may be precarious. Thus in northern forests of Canada the snowshare hare is preyed on by the Canada lynx. The lynx depends heavily on hares when they are abundant and the hare has the capacity to outbreed the lynx. Although other species such as horned owls also prey on hares, there are not enough predators to control the increase of the prolific hares. When plant foods are plentiful for hares, the hares increase to great numbers. Under these circumstances, their predators can increase also to high levels. But finally the hares become more abundant than their food supply, or reach a level where the combination of food shortage and stress, resulting from excessive social interaction between overcrowded hares, causes them to die off in great numbers. When this happens a formerly abundant food source for predators will disappear over the course of a winter and spring. Since there are few alternative kinds of prey animals present in any abundance, the predators either die of starvation or are forced to migrate to other areas where food can be found. When predators are once again scarce, and plant foods recover, the hares once again increase. For reasons still somewhat obscure, this increase and decline in hare and lynx populations follows a ten-year cycle, with peaks being reached every nine or ten years. Fur records maintained by the Hudson Bay Company in Canada have enabled scientists to trace this cycle back for over two centuries.

Such cyclic increases and declines are characteristic of the simplified ecosystems of the Arctic. One of the best known is the three- to four-year cycle of the lemming, an Arctic mouse that reaches exceedingly high levels of abundance at the peak of the cycle and may scarcely be found after the crash.

By contrast, in more complex and diversified communities there are usually many predators capable of feeding on any form of prey and many alternative prey species available to predators when any one species becomes scarce. Consequently, it is unusual for any one species to become excessively abundant and equally unusual for it to decline in numbers to an extremely low level. Although some populations do fluctuate in abundance, this occurrence is far less usual and less dramatic than those of the simplified communities of cold or arid lands where such occurrences are the rule rather than the exception.

Unfortunately, human use of ecosystems usually involves some degree of simplification, from a diversified to a less complex situation. Seldom are all of the species in a biotic community equally useful to human purposes, and it is a normal tendency for man to favour those species most useful to him. This has been the basis for agriculture and animal husbandry and, more recently, for forestry. So long as this simplification is not too extreme or widespread, the natural checks and balances remain operative. However, as man's efforts to obtain greater yields from larger areas have intensified, there has been a growing tendency to devote more and more land to monoculture, the continued cultivation of a single crop. This creates, artificially, a more severe imbalance than occurs naturally in arctic or desert ecosystems. Severe fluctuations in those predators (or parasites) that will feed upon this single crop become normal and, as a result, a constant war against agricultural pests must be waged. Where this in turn leads to further simplification, such as occurred where DDT is used continuously in cotton fields – and has eliminated the natural predators of cotton pests – the situation grows more extreme, forcing either abandonment of the monoculture or development of means for pest control that favour continuation of natural balances.

There is growing evidence of the value of maintaining natural diversity, both as a means for maintaining community stability and for other more direct benefits to man. In particular, since each species represents a unique array of genetic materials, each has a potential value to man's welfare that is only revealed when needs arise and the necessary studies are carried out. Some may contain biochemical substances of value: the quinine derived from the cinchona plant is an example, and the discovery of antibiotics was dependent on the existence of the naturally-occurring *Penicillium* mould. Knowledge of the Rh factor in blood came from the study of a distant relative of man – the Rhesus monkey. All of these advances depended on the existence of wild species, often of species considered to have no particular value or, in the case of *Penicillium*, to be a nuisance. Each year many more species make their contributions to medical science, agriculture or other human purposes.

Perhaps even more important to man in the long run is the psychological value of having available to him the great diversity of natural communities from which his own human existence has been derived. It is difficult to measure such a value, yet few would deny that, when given the opportunity, people seek contact with wild or domesticated nature and gain psychological benefit from such experiences. Finally, as we have seen, the sum total of functioning life in the biosphere makes possible the continued existence of any one species, including man. We reduce this diversity of life, therefore, only at our peril, since we do not yet know how all of the parts function and which are essential to the whole.

Chapter 3
The Impact of Man:
The Historical Record

We do not know and perhaps never will know where the human species appeared on Earth nor when this occurred. Africa has the longest known record of inhabitation by manlike primates and it seems highly probable that somewhere in tropical Africa the human race had its beginnings. Just where and when the present species, *Homo sapiens*, appeared is also uncertain but, once again, Africa seems the most likely continent. Although we can only speculate on the early relationships between man and his environment, there is much to be learnt from the study of contemporary groups of peoples on Earth. It would appear that virtually every type of culture which the human species has devised during the course of evolution is still preserved, albeit in a modified state, in some existing population.

In the great centres of technological civilization it is sometimes difficult to realize that there are people in the deserts of Australia and southern Africa who are still existing at the hunter–gatherer stage of existence, largely unaware of the technological world. In the forests of the Amazon and the Congo are those who combine this form of existence with some reliance on a primitive subsistence agriculture. Elsewhere are those who exist as peasant farmers little different in their ways of life from those who tilled the fields of the Middle East or planted their crops in forest clearings in South-East Asia 3000 years before Christ. In the Sahara and the deserts of Arabia are nomadic pastoralists following age-old patterns of dependence between man and his domestic beasts. The world of technology, rapid transport and instant communication still has not swept all of these peoples into its net. There still remain other ways of living with the environment. From these there is much that can still be learnt.

From the evidence available we can see that the effects of man upon the biosphere have, for the most part and during most of human history, been relatively benign. Yet wherever man has lived for long he has changed his environment. Always he has attempted to make it more suitable for his continued existence. Often he has succeeded.

In the following pages the effects of some of these primitive technologies will be examined.

Fire

One of man's oldest techniques has been his use of fire to modify and shape his environment. Wherever vegetation will grow in sufficient quantity to carry a fire, and becomes dry enough to burn, we find evidence of the persistent occurrence of fire. Arid deserts will not burn, nor will the more humid tropical forests, nor barren mountain tops. Most other areas will burn, and the natural frequency of fires seems determined in part by the rate at which dry vegetation accumulates on the ground. Lightning is a prevalent factor in the atmosphere and eventually will serve to ignite fuel. But the frequency with which fires occur has been increased by man's activity. Fire can be used to drive game and thus make these animals more readily available to the hunter. It can be used to kill some kinds of animals, the cooked meat of which then becomes available to the human scavenger. It can be used to modify vegetation and make it more attractive to these animals that man liked to hunt or, later in human evolution, to those domestic animals that he depended on for milk, or meat, or hides.

The replacement of woodland by savanna, the replacement of fire-susceptible plants by those resistant to fire, has gone on in many parts of the Earth over tens of thousands of years. However, so long as human numbers were limited and the space available for human activities was relatively large it could not be said that the effects of fire were harmful or destructive. It is only since human numbers have greatly increased and the pressures upon all lands have intensified that fire has become a destructive force on Earth.

Today the misuse of fire destroys in each year millions of hectares of valuable timber resources that would otherwise contribute useful products for man. Burning of vegetation at the wrong season of year or at too frequent intervals can remove valuable forage resources and lead to the replacement of highly valued plant communities by others of lesser value. Misused, fire can cause accelerated erosion of the soil and lead to the long-term loss of productivity. In more arid lands, fire can bring about the extension of deserts into formerly productive steppe or savanna. Also, and not least important, the prevalence of wild land fires is a contributor to air pollution, and the accumulation of smoke particles in the atmosphere, along with dust and ash, and materials from urban chimneys and motor car exhausts, may lead to adverse climatic effects.

Nevertheless, fire remains an important tool to be used in forestry, range management and wildlife management. It must, however, be used skilfully and carefully managed to produce the results desired. We do not as yet have all of the necessary knowledge and skills to make the most effective use of fire. We have not yet brought together and analysed the knowledge that is available in the various biomes in all of the different nations of the world. As we see the continued misuse of fire, the need for accumulating the knowledge required for its proper use becomes increasingly apparent.

Hunting

There is considerable disagreement about the role of primitive man in the post-Pleistocene extinctions of large mammals. It has been pointed out that many of the species of large mammals became extinct at a time when highly effective hunting cultures had been developed, that many became extinct at approximately the same time in history, and that there are no obvious causes for these extinctions other than the activities of man. In North America, the horse, camel, large-horned bison, mammoth, dire wolf, ground sloth, sabre-toothed cat, and many other species became extinct. In other continents similar lists of post-Pleistocene extinctions may be compiled. However, others find it difficult to believe that relatively small numbers of hunters, armed only with primitive techniques, could have such devastating effects, and cite evidence from modern game-control efforts to show the resilience of wild animal populations exposed to intensive hunting pressures.

Since it is not possible to separate the effects of habitat change from the effects of hunting nor to measure the precise role of either in the distant past, we can only leave it as a possibility that primitive man played an effective role in the extermination of many species of animals. For some the evidence is conclusive; the extinction of the New Zealand moas by the first human residents of these islands is an example.

There is no doubt at all that man during historical times has been an important if not a major factor in bringing about the extinction of animal species. In Europe, for example, the aurochs and lion have been completely exterminated since Roman times. The wolf, bear, bison, lynx, chamois, ibex and most other large mammals have been greatly reduced in numbers and restricted in their distribution to a few wild or protected areas.

Yet, except in the most intensely settled areas, the more surprising fact is not the fact of extinctions, but that of survival. In North America, for example, despite the enormous and unrestricted pressure of hunting directed at the larger mammals, including the trapping and poisoning of many species – no species has yet become extinct. Many have been locally exterminated, and some races have been completely destroyed, but the species have survived in a wild state. Among birds, the great auk, the passenger pigeon, the Carolina paroquet and the Labrador duck have been exterminated, but all other species have survived, although local races have vanished. Elsewhere, admittedly, the record has not been as good. On islands, in particular, the number of species exterminated is large, but these represent restricted and highly vulnerable populations.

Most hunting by primitive men had best be viewed as a normal form of predation on wild animals. Species became adapted to it just as they have adapted to the presence of other carnivores or natural enemies, and man and wildlife lived in a fluctuating balance.

Only in comparatively recent times has man become sufficiently numerous and powerfully armed to be certainly capable of major exterminations of wild species through hunting. In areas where hunting pressure cannot be controlled through laws and regulations, or through parks and reserves, the total extermination of many species appears imminent.

For many people hunting is an enjoyable form of recreation. Properly conducted it can be beneficial to the management of wild animal populations as well as a source of protein and of sport for man. The commercialization of hunting in some areas where wildlife is naturally abundant offers a prospect for yielding, on a sustained basis, supplies of meat and other animal products comparable or superior to those that might be obtained from the same areas through the use of domestic livestock. Although this prospect has been realized in some areas, there is need for much more study, particularly since the risk of hastening the extermination of certain species is involved wherever hunting and marketing of wild animals is not strictly controlled.

Fishing

Like hunting, fishing has been a long-standing technique by which man has obtained supplies of protein from his environment. Originally fishing was conducted on much the same basis as hunting. However, when man began to practise agriculture and later to develop civilization, wild animals tended to vanish from his immediate environment. Fish remained, however, in streams and lakes, and in the ocean. The means for exploiting them more intensively were developed in early historic times, so that fish became a staple source of protein for peoples living near potential fishing grounds. Although fishing was originally confined to near-shore areas, the development of more efficient vessels and exploitation techniques has permitted mankind to extend his fishing pressures throughout the world's oceans and to draw upon a wide range of marine life for his support.

Because of their high fertility and consequent capacity to recover from loss, it has been much more difficult to exterminate fish species than species of terrestrial animals. However, where a species is restricted to an accessible and limited habitat, this rule does not apply. The greatest number of exterminations have occurred among species in confined inland waters. In the oceans, species exterminations are much more unusual, and those species that are known to be extinct represent either slow-breeding marine mammals, or birds which are dependent upon land for part of their life cycle.

Although species do not readily disappear in the marine

environment, fisheries become depleted to a stage where their commercial value is eliminated or drastically reduced. The California sardine and Atlantic salmon are examples of marine fisheries that have undergone drastic reductions and, in the former instance, show no signs of recovering past abundance. Marine mammals, the whales and seals, have been particularly hard hit by heavy exploitation. The right whale and blue whale were near to extermination before effective protection was provided. The Caribbean and Mediterranean monk seals have been reduced to near extinction.

Under proper management, fisheries may continue to yield high and increasing amounts of food for man's use. The potential fisheries yield from the oceans has been much debated, but it is evidently more than double the present yield. Unfortunately, proper management requires a level of international agreement that has not been easy to obtain. A few examples of highly successful treaties are available – those governing the Pacific northern fur seal and the Pacific halibut are illustrative – but over most fisheries there is as yet no international control.

Food gathering

Opinions differ on the extent to which early man was dependent upon animals that he killed as compared to plant foods that he collected from the woods and fields. Some see early cultures as being centred around hunting and envision early man as a superior predator, others believe that he had a mixed dependence upon those animal foods that were easily obtained and the plants that were more generally available. There is no doubt that at various times and places during the Old Stone Age, the Palaeolithic, highly efficient hunting cultures came into being. There seems little doubt, also, that even the most effective primitive hunters make wide use of a variety of plant foods and those animal foods that are gathered rather than hunted – the grubs or larvae of certain insects, termites, various molluscs, birds' eggs and the helpless young of certain mammals among them. Food collecting depends for its success on the ability to gather in, and to some extent store, the more nutritious plant parts – seeds, fruits, bulbs, tubers – in which the reserve of carbohydrates, oils and proteins are stored. There is no reason to assume that man's food gathering had any more marked effect upon his environment than the similar activities of other mammals with similar habits. It seems most likely, however, that plant collecting, and the storage of plant foods that became essential when people first occupied the seasonally cold or seasonally dry lands, led logically to the development of agriculture. Stored bulbs, tubers or nuts would sprout and grow, and old camp sites, when revisited, might yield an increasing supply of the valuable food plants upon which people had learned to depend.

With the increasing importance of agriculture, interest in wild foods diminished, but it has never disappeared. People still value many of the plant products of the wild, even where other foods are abundant, and they will expend great amounts of energy in a search for wild fruits or berries, fungi or the honey from wild bees. Where agriculture yields only marginal support for people, the role of wild products increases in importance, and indeed makes it difficult to evaluate the nutritional adequacy of the diets of many primitive people.

Plant domestication

There came a time in the prehistory of mankind when, in certain places and among certain people, a shift was made from a dependence upon wild foods to a reliance upon the tame. This change, since it took place during the years that archaeologists have categorized as the New Stone Age, has been known as the Neolithic Revolution. There is little doubt that it had revolutionary effects upon those affected by it, and led in time to the development of civilization.

There seems no reason to believe that agriculture had a single point of origin, nor that it derived initially from the domestication of any one group of plants. Nevertheless, the spread of agriculture and of domesticated plants has been traced from certain centres in which its success was most outstanding. Carl Sauer, a geographer noted for his studies of plant and animal domestication, believes in a tropical centre of origin for agriculture in both the Old World and the New World. He believes that the plants that grow from bulbs or tubers, or that reproduce vegetatively from stem cuttings or shoots, were the first to be domesticated and that the early agriculturalists were planters rather than seed sowers. He has described the dissemination of the agricultural idea into areas where the original domesticated plants did not grow well, the drier margins of the tropics. In this region he sees the substitution of seed crops for the root crops of the tropics. Seed-crop agriculture, based primarily on annual grasses – the cereal grains – spread then from a south-west Asian and north-east African centre throughout the Eurasian and African land masses. Root-crop agriculture spread from its centre in south-east Asia through the tropical Pacific and westward into tropical Africa. In the Americas, a similar diffusion spread out from centres in northern South America and Mexico.

The impact of agriculture upon the environment was, in time, to become enormous. At first, however, its greatest effect was in permitting greater numbers of people to remain for longer periods of time within a single area. The shifting camp of hunters and food gatherers gave way to the permanently established village depending on the continued yield of cultivated fields.

Primitive agriculture took two different directions. One, towards the permanent garden or field, was dependent for its success either upon deep and fertile and highly stable soils, or upon the continued care and fertilization through plant and animal manures of less fertile, but relatively stable, soil areas. This led in time to one of its most successful and productive offshoots, the development of irrigation agriculture on the rich alluvial soils of river bottom lands. The other direction, particularly characteristic of the less fertile forest soils, was the pattern of shifting agriculture – variously known as 'slash and burn', '*milpa*', '*ladang*', '*chitemene*' and so forth.

So long as either of these systems was used to support a sparse population of people in a localized area, little harm to the environment could take place. Soils that became eroded or depleted of nutrients could be allowed to recover under natural vegetation. The new agricultural landscape of fields and gardens subtracted nothing from the much more extensive area of wild and natural lands, but added an additional element of diversity which certain wild species – later to be known as animal pests and weeds – found attractive.

With the increase of human numbers, with the reorganization of human society that accompanied the rise of civilization, with the growing demand for crops to meet the needs of urban populations and to enter into commerce between centres of population, agriculture was sometimes forced from a benign relationship with its environment to one of conflict. Particularly under the exigencies of war and conquest, land care suffered. Wind and rain took their toll in soil erosion from fields left bare. Hillside terraces fell into ruin and their soil disappeared in growing gullies. Some lands were cultivated, in a desperate effort to grow subsistence crops, until their fertility was exhausted.

In all the old homelands of agriculture one finds evidence of land abuse and deterioration – fantastic patterns of erosion where once were upland fields in northern China, deserts covering formerly productive lands in the Middle East and North Africa, scrub jungle covering the ruined cities of the Mayans, barren and eroded slopes where hillside farms once provided support for the Toltec centres. The collapse of many ancient civilizations has been attributed to failure in land-use practices. There seems little doubt that it was a contributing cause. The ruins of man's great achievements of the past have fulfilled their role as warnings for the future.

In the humid and subhumid tropics, where shifting agriculture has prevailed, growing numbers of people have placed undue pressure upon the lands. The success of this agriculture on tropical forest soils was based upon a long period of forest fallow following on a brief period of forest clearing and cultivation. During the long fallow, natural vegetation would rebuild the soil and restore its nutritional elements. During the short cultivation, this fertility could be drawn on for man's support. But the capacity of such systems is limited. When asked to support too many, the fallow time must be shortened and the cropping time extended. Soils have little time to recover from use. Natural vegetation loses ground. Soil depletion and soil erosion take their toll. Some tropical soils readily become laterized, covered by a hard ironstone crust, when subjected to clearing and agricultural use. Such soils may be permanently ruined for further use. Lands that could have supported a few people almost forever, become unable to sustain great numbers for more than a few years. New lands must then be sought and cleared, but the availability of new lands is constantly diminished.

Irrigation

In the lands where man first practised irrigation as a means for enhancing agricultural yield, the rivers rose in flood-time each year following the period of heavy rains in their headwaters. After the flood peak the waters receded, leaving behind banks of sand, silt or mud in which the seeds of annual plants would germinate and grow. No doubt pre-agricultural man visited these flood plains during his hunting and fishing expeditions and harvested some of the more useful plants. It would be a logical place also for him to experiment with the growing of seed plants. In time, as agricultural practices developed, villages would grow and human numbers increase. The need for additional land to grow crops, the obvious value of river water to keep the soils moist, and the human inclination to play with water, all would lead to early attempts at irrigation. Along the Nile, the Tigris–Euphrates and the Indus, these experiments proved profitable. The channelling of river water and its use to flood the alluvial soils made possible the growing of crops in areas well removed from the river's edge and permitted full advantage to be taken of the warm temperatures and long growing seasons of these subtropical environments.

Primitive irrigationists took advantage of the seasonal elevation of the rivers to lead their waters through canals into lands left dry by the retreat of the normal annual flood. Later, with the aid of simple mechanical devices, water was elevated from the river to flow through canals built on higher ground. The soils so watered were virtually inexhaustible, since each year they were renewed by the thin deposit of silt carried by the river from sources farther upstream. The drain in nutrients taken by the annual crops was balanced by the addition of nutrients carried by the river. Crops could be grown in excess of local needs, and the surpluses could be used for the support of civilizations and, in time, empires.

The success of the system was dependent on the stability of the river flow, on enough but not too much water being available. This led, as populations grew to high levels, to years of famine when the river flow was low and the upper croplands could not be irrigated, and to years of plague when the river was

Plate I The influence of man reaches throughout the biosphere: even the penguins of Antartica are affected

Plates 2 and 3
left and below
The byproducts
of our
technology
destroy the life
of our planet and
desecrate
the land

Plate 4 *right*
Hillside stripped
of every scrap of
vegetation forms
a 'tree of erosion'
pattern in this
photograph taken
from a thousand
metres up

Plate 6 The oldest technique for land management – fire. Through burning, primitive man changed the landscape from woodland to grass

Plate 7 A Masai goatherd tends his flock. Even here, the grazing lands are overstocked and nature's balance is threatened

Plates 8 and 9 The rise of industrialization changed the face of the earth –
old landscapes were crowded out and new landscapes created

Plates 12 and 13 Pittsburgh (the 'Smokey City') used to be one of the blackest cities in the USA. City laws against smoke, wider use of natural gas, fuel oil and electricity, and more efficient coal burners have made Pittsburgh a cleaner place

Plate 14 The movement towards conservation: National parks are now established in many countries of the world

Plate 15 The movement towards conservation: Steps have been taken to save the Galapagos Islands with their unique wildlife

too high and the lower crop lands were left flooded, marshy and steaming – a breeding ground for malarial mosquitoes and other disease-carrying insects. Yet the Nile flow was usually dependable. Civilization thrived, despite the rise and fall of dynasties and the presence of invading armies. On the Tigris–Euphrates, a different situation prevailed. The continual abuse of headwater lands, through excessive farming pressure, uncontrolled numbers of livestock and the frequent passage of invading armies, led to increasing instability in river flow, and ever-growing loads of silt moving down the river – this blocked canals and buried fields under greater loads of debris than could be handled.

The success of the system was also dependent on an abundance of water and good drainage. Where water was too scarce or drainage impeded, salts moved upward through the soil following the capillary pull from surface evaporation. Soil surfaces in time became too saline or alkaline for the production of most crop plants. Formerly productive lands were forced out of production. In the Indus region and Mesopotamia, the early civilizations collapsed. Only Egypt has remained productive up to recent times.

Most of the great civilizations of early times have grown up in warm, seasonally dry or arid lands, on the basis of crop surpluses built up from irrigation agriculture. The term 'hydraulic' civilizations has been used to describe them. The Romans are noted for their skill in hydraulic engineering. Going far beyond any developments carried out in Ancient Egypt, they led water from mountain streams through flumes and aqueducts to irrigate distant areas. In Ceylon, an elaborate civilization developed in the drier half of the island through use of reservoirs constructed in the high-rainfall areas of the mountains, and a canal system that carried the water for use in the dry lowlands. Similarly, in Peru the Incas developed an elaborate irrigation system that permitted the use of soils in the lowland deserts. In South-East Asia the use of the fertile flood plains of the great rivers permitted the support of millions of people and thriving cultures.

Apart from the salinization problem that has been noted and the encouragement given to waterborne diseases, the environmental effects of early irrigation agriculture were on the positive side. Although the removal of much riverine vegetation and animal life was necessary, the total effect was an enhancement of productivity and diversity. Deserts were made to bloom, but at that time there was always a surplus of natural desert.

Forest exploitation

Perhaps the first useful thing that mankind discovered about wood was its ability to burn. Today, despite millenia of civilized life and the discovery of the means for using the energy of the atom to heat the fires of the home, the principal use for wood, over most of the world, is for fuel. Forests remain an important source of firewood for cooking and space heating, and a source of charcoal – a lightweight and readily transported fuel.

Pre-agricultural man made little use of forests. His fires and clearing encroached upon them, but his total impact was confined to the dry forests on the edges of grassland and desert. Here he had his greatest effect. Closed woodland disappeared from great areas to be replaced by the fire-adapted savanna.

Post-agricultural man made a greater impact, drawing on the woods near his villages for fuel and building materials. Those who lived by the edge of the rivers, lakes and seas learnt initially that logs float and that hollowed logs could carry people in comfort. The idea of boats developed and so also the need grew for wood for boats and, in time, ships' timbers. But civilization and its technologies brought more pressure on the forests. We are told how the Israelites drew upon the cedars of Lebanon to build the temple of Solomon. But the same timbers went for many purposes to Mesopotamia, Egypt, Crete and Turkey, and the Lebanon forests eventually disappeared, their seedlings browsed by goats and the soils that supported them eroded away. In northern China great areas of forest disappeared to fuel the kilns of the makers of pottery and porcelain. Erosion followed in the wake of deforestation. In the Valley of Mexico, the boundaries of the forest were pushed back in the search for fuel to keep the Toltecs and the Aztecs warm and to cook their food. All around the Mediterranean, where civilization reached an advanced level, forests vanished.

The impact of pre-industrial man, however, was localized. Forests for the most part were less used than cleared. The effort to win new land for farm or pasture pushed the forests farther away from the settled land. But the balance was precarious. During the Dark Ages in Europe, and again following warfare and plague in medieval times, forests came back, taking over the fields and crop lands and moving in around deserted villages. Nevertheless, on soils well suited to agriculture, the removal of forests became permanent and for the most part they were replaced by stable and productive farming lands. This conversion of forest to farmland in western Europe permitted, through providing agricultural abundance, the development of industrial civilization.

Industry and machines changed the balance. To fire the furnaces of England the forests disappeared and the development of coal resources became essential. To meet the various wants and needs of advanced culture, a selective search for quality timbers took place – sandalwood, cedarwood, mahogany, ebony and other choice woods were hunted down wherever they grew. Early industrial civilization seemed an enemy of forests. But the more serious attack on woody plants took place, not in the well-watered lands where forests grew, but in the deserts and semi-arid regions where vegetation maintained a precarious balance with its environment. There, the demand for wood by pastoral peoples led to the removal of the scant brushy cover. Lands that had once supported their own peculiar kind of drought-adapted life became real wastelands indeed.

Animal domestication

The increase and spread of human biomass has been catastrophic for most species of animals on Earth. For a select few, however, it has brought a form of success. There are perhaps as many domestic mammals on Earth as there are people. When domestic fowl are included, the numbers of animals under man's special care greatly exceeds human numbers. The impact of pre-industrial man alone on his environment was significant but hardly of major importance in the biosphere. The impact of man plus his domestic animals has been of a different order of magnitude. Together they have modified great areas of the Earth and too often have brought ruin to once productive lands.

The domestication of animals, like that of plants, dates back to the Neolithic. Where two streams of plant domestication can be traced, that of the tropical planters and of the dry-land seed sowers, so also can two lines of animal domestication. The tropical forest peoples have cultivated the household animals – pigs, dogs and jungle fowl. The seed sowers have cultivated the species that, like themselves, depended upon the grass of the Earth, the herding animals, grazers and browsers. The effect of the tropical forest animals upon the environment has been minor, that of the herding grazers of major consequence.

It is generally agreed that the effective domestication of animals, as opposed to the keeping of wild pets, had its origin among agricultural peoples. These, unlike the hunters, had both the time to experiment and an interest in the taming and breeding of other living things. In the lands between the Caucasus and the Indus, and from Ethiopia to the southern fringes of Europe, the principal herding animals were brought under human tutelage – cow, sheep, goat, dromedary, camel, donkey and horse. Elsewhere, in a few other parts of the world, a few other mammals were domesticated – the llama and alpaca in the Andes, the yak and bactrian camel in the central Asian highlands, the water buffalo and elephant in South-East Asia. There have been other animals domesticated temporarily, the African elephant among them, but the animals that have been kept in permanent service have been few in species, although they were to become very rich in domesticated breeds and varieties. Thus, there are far more distinctive and unusual breeds of domestic dogs than there are of all the wild canids that have ever been known and, were some of these to be encountered as free-living wild animals, they would certainly have been classified by early naturalists as distinctive genera if not families of animals.

The practical, commercial orientation of those who live in the technological world leads to a consideration of domestic animals as so many unit suppliers of meat, eggs, milk, wool or other products capable of producing a particular level of monetary income. But this is not the view of the pastoral peoples. To those who domesticated animals and to the pastoral peoples who are their linear descendants, the relation between man and his flocks or herds is not so simple. There is a symbiosis between man and animal not to be quantified in simple economics, the roots run into the areas of religion and sociology. One had just as well

measure the meat producing capacity of the house cats of Paris as calculate that of cows of the traditional Samburu pastoralists. The figures obtained in either case would be irrelevant, and therein lies part of the problem.

The care of domestic herding animals during Neolithic times developed along two separate lines. Some were kept in and near the village, and were fed partly on the stubble and wastes from the agricultural fields, and partly on nearby pastures. These were the animals particularly used for the production of milk and its products. But in time, in every area, a surplus would be produced beyond that which the fields and meadows could support, and this must be taken farther afield. Those who cared for the cattle were at first associated with the farming lands. But, eventually, the need for further pastures led to an increasingly remote and independent life. Somewhere in history the separation became almost complete, and the nomadic pastoralist came into existence. These depended largely on their flocks and herds for their sustenance. Agriculture became an incidental thing in their existence, or they learnt to trade with settled peoples for the things that they required. Like the wild grazing animals that they tended to displace, the nomad and his herds developed an adaptation to the ecology of the steppes and desert, the dry savanna and the mountain pasture. The secret of success lay in movement, in travel to follow the rains or to move upward in the mountains in pursuit of the springtime growth of new pastures. In the savannas also the use of fire to remove old leafage and promote new growth enabled the herds to feed well and to thrive.

Wild grazing animals tend to move freely and feed selectively. This prevents excessive grazing pressure upon the vegetation of any one area and allows animals to pick the more palatable and nutritious growth produced by each plant. Where numbers are not excessive and the variety of species is great, a relatively uniform use of wild vegetation will occur, similar to that provided by domestic stock on planted, well-managed pastures. Nomadic pastoralists developed a similar technique. Where their movements were not interfered with and their numbers did not become excessive, their total environmental effect would bc in no way destructive. But these conditions did not always prevail.

The settled ways of civilization and the military needs of empires do not fit in well with the free-roving lives of nomadic peoples. Nomads who are confined and whose herds are allowed to increase in a limited space become a destructive force. Excessive grazing and the trampling of too many hooves destroy the vegetation and compact the soil. Grasslands are changed to deserts, deserts to wastelands.

To the farmer, land that is not settled and ploughed is regarded as empty land suited to his occupancy. To the nomad, such land may be a choice pasture held in reserve for winter or dry-season use. When he returns from his seasonal travels to find his pasture ploughed, he is likely to take retribution. Thus, the seeds of conflict are readily sown. The conflict between nomad and settled folk has raged throughout history. The last, Earth-shaking surge of nomadic peoples was that led by Tamerlane and brought a finish to many a long-established city culture. But even today the old battles continue in a lesser form, and the old antagonisms add fuel to more modern territorial conflicts.

What were once the world's prime grazing lands are now ploughed under and growing domesticated plants – the black earth prairies of the Ukraine, the deep rich soils of the American middle west, the pampas of the Argentine. Pastoralism has been pushed back to the marginal lands where seasonal drought or total scarcity of precipitation limit the spread of agriculture. In this more limited area the old nomadic ways are less successful, the dangers of land destruction are greater.

Throughout most of the grazing lands of the world, serious damage to the vegetation and soil has occurred. In some places where modern techniques of wild-land management based on ecological principles have been applied, the damage has been repaired. But in much of the underdeveloped world, and in far too large an area of the technologically advanced countries, the old destructive usage of resources continues. Too often efforts to improve the lot of pastoral peoples have led to greater and more accelerated damage. The concept of carrying capacity and the related idea of sustained yield have not yet been well understood.

Pollution

Pollution may be defined as 'the addition to the environment, at a rate faster than the environment can accommodate it, of a substance or a form of energy (heat, sound, radioactivity, etc.) that is potentially harmful to life'. Pollution is therefore neither a new nor a man-made process. The natural accumulation of toxic salts in great quantities has made the barren salt pans and playas of arid and semi-arid lands, has created the salt lakes and dead seas that are uninhabitable by most living things. The excessive build-up of organic wastes under anaerobic conditions formed the oil, coal and gas deposits on which we rely today. One of the great sources of air pollution is volcanic activity, even today this can dwarf any contribution mankind is capable of making to the total sum of particulate matter in the upper atmosphere. Eutrophication, the enrichment of water bodies by nutrients, is an age-old process and has been in part responsible for the development of the swamps and marshlands of the Earth, as well as the productive lakes and estuaries.

It has been said that man is by his very nature a polluter, but this is not necessarily true. So long as human numbers were small, and populations dispersed and mobile, nature was fully capable of assimilating and processing human waste products. Only when populations increased and people settled for long periods in limited areas did man-made pollution begin. We can still see evidence of 'solid waste' pollution in the shell mounds or kitchen middens that mark the long-established camp sites of primitive man. In such areas, during the periods when they were inhabited, there were undoubtedly problems of disposal of other human wastes as well, and perhaps localized problems of water pollution. Unquestionably, when man began using fire on a large scale he contributed substantially to air pollution. Natural or man-caused forest, brush and grass fires are still a major contributor of pollutants to the atmosphere. Nevertheless, pollution in the accelerated form which concerns us today, is basically a problem of civilization. As such, it became a local problem in the cities of ancient and medieval times. Only recently, however, has it become a global problem.

Lewis Mumford has given us some idea of the pollution

problems that beset ancient Rome. In his words, 'surely it is no accident that the oldest monument of Roman engineering is the Cloaca Maxima, the great sewer, constructed in the sixth century (BC) on a scale so gigantic that either its builders must have clairvoyantly seen, at the earliest moment, that this heap of villages would become a metropolis of a million inhabitants, or else they must have taken for granted that the chief business and ultimate end of life is the physiological process of evacuation.' The great sewer, useful though it was for removing waste water and human sewage from those areas that it serviced, simply deposited these wastes in the receiving waterway and ultimately the Tiber estuaries. Although there are no accounts of the effects, they must have represented the most severe water pollution problem of that period of history and no doubt contributed to the periodic plagues that ravaged the Roman population.

However, it was in the disposal of other types of refuse that Rome, in Mumford's words:

records a low point in sanitation and hygiene that more primitive communities never descended to. The most elementary precautions against disease were lacking in the disposal of the great mass of refuse and garbage that accumulates in a big city .... If the disposal of faecal matter in carts and in open trenches was a hygienic misdemeanour, what shall one say of the disposal of other forms of offal and more ordure in open pits? Not least, the indiscriminate dumping of human corpses into such noisome holes, scattered on the outskirts of the city, forming as it were a *cordon malsanitaire*.

The Italian archaeologist, Rodolfo Lanciani, had the misfortune to excavate one of these refuse pits, a *carnarium*, some two thousand years after it had been filled and abandoned by the Romans. He was, in his words, 'obliged to relieve my gang of workmen from time to time, because the stench from that putrid mound, turned up after a lapse of twenty centuries, was unbearable, even for men inured to every kind of hardship, as were my excavators.' Clearly, pollution is no new problem.

With the first beginnings of the industrial revolution, a new aspect of pollution took form, that of air pollution resulting from the burning of fossil fuels. Eugene Ayres has pointed out that Edward I of England, in the fourteenth century, found it necessary to ban the burning of coal in London, under penalty of death, because of his concern that 'the health of the Knights of the Shire should suffer during their residence in London.' But the ban against such burning was to be lifted and London's reputation for polluted air grew until recent times. In all of the great metropolitan centres of the world pollution problems grew increasingly severe. Friedrich Engels, seeking information on the condition of the proletariat in Europe, has left disturbing pictures of the pollution that existed in Lancashire during the early nineteenth century. Describing the River Irk, he states,

At the bottom flows, or rather stagnates, the Irk, a narrow, coal-black, foul-smelling stream full of debris and refuse which it deposits on the shallower right bank. In dry weather, a long string of the most disgusting, blackish-green, slime pools are left standing on this bank, from the depths of which bubbles of miasmatic gas constantly arise and give forth a stench unendurable even on the bridge forty or fifty feet above the surface of the stream.

Considering such descriptions, which could be matched anywhere in the urbanized world at that time, one may wonder why only today the problem of pollution is attracting major attention. In fact, urban pollution did attract major attention in the nineteenth century. Following on investigations such as those of Friedrich Engels, city governments, for the first time since the days of ancient Rome, began to accept their responsibility for the provision of sanitary waste disposal. Sanitary sewers were constructed in major cities of Europe and the United States, starting particularly in the 1840s. Earlier sewer systems were designed for carrying storm run-off and not human wastes. Around the same period, the ravages of cholera and typhoid finally forced recognition of the need to separate drinking water from waste water, and urban authorities began to take responsibility for the supply of potable water to their communities. Nevertheless, until relatively recent times pollution was a localized problem. Large cities were uncommon, most of the land was agricultural or wild. Within the cities themselves those most affected by pollution were those with the least power and influence to do anything about it. Those who made decisions for a nation were usually well-shielded from ugly sights, sounds or smells.

Chapter 4
The Impact of Man:
The Recent Record

It is difficult to draw a line anywhere in human history that separates old from new. Change is continuous. Even our most recent technological gains have their roots far in the past, whereas some of our oldest practices are only now having their full effect. We have examined some of the pollution problems that beset the ancient world. We can also find, far in the past, a similar array of difficulties associated with other aspects of urbanization, industrialization, transportation, major engineering developments and all of the other processes and activities that we consider characteristically modern. Ancient Rome even had its 'new towns' programmes intended to solve the problem of excessive growth in and around the capital. As Lewis Mumford has pointed out, these were no more successful in achieving this goal than their modern equivalents around London and Paris. Nevertheless, it would be simply argumentative to maintain that there is not a quantitative and qualitative difference between the world of even fifty years ago and that of today. The environmental problems of today are of a magnitude and a scope that was not approached or equalled in earlier times. To use an analogy, the environmental problems of the past century and before were like a number of small, slowly burning grass fires. Each could have been approached by the people most directly affected and then beaten down or suppressed by a variety of means. But they were not. Today they have all run together and we are faced with a global conflagration. Only by a unified effort and the use of the most effective techniques can we hope to bring it under control.

Industrialization and its consequences

Although it is misleading to think of an Industrial Revolution as an event occurring in the nineteenth century and thereafter transforming the ways of life for people throughout the world, there is nevertheless some reason for the general acceptance of this idea. It must be admitted that all of the factors which acted together to produce the sudden surge of industrialization had their origins centuries or millenia before, but there is little doubt that during the nineteenth century, more than at any preceding time, they came together to produce spectacular results. In the centres of industrial activity the old rural–agricultural way of life gave way to an urban–industrial way of living. The use of fossil fuels, coal and later petroleum and natural gas, began to dominate, and other fuels or energy sources retreated to minor roles. More and more, individuals were oriented towards goals of quantitative production, often meaning that their working activities were channelled in ways that permitted them little satisfaction in achieving these goals. The factory assembly line, in which each worker played a minor role and none was responsible for the end product, replaced the old village workshop or the craft guild, in which each felt a personal responsibility for the quality of the thing that was being manufactured. Admittedly this feeling of interest in the quality of the end product has not been entirely lost in the modern factory, or we would not be able to rely upon the performance of the machines and materials produced to the extent that we do. But for most workers the satisfaction that comes from having done a good day's work well is now difficult to achieve.

The environmental impact of industrialization was manifold. First, it brought a demand for ever increasing amounts of raw materials. This led, in many areas, to ruthless exploitation of natural resources with little or no regard for environmental consequences. The destruction of the forests of many areas of the world, with no thought for sustained production, was one such consequence. The ugly scars left by mining on many of the landscapes of the Earth is another. This was particularly severe where mineral resources of value were available at or near the surface of the ground. Essentially the entire skin was removed from the Earth in such areas; the minerals were removed, and nothing was restored. The ugly scars from strip mining of coal, with all of their consequences in stream erosion, silting and pollution, are to be seen in the eastern United States as well as other areas of the world. In order to achieve the immediate profits to be obtained from their subsurface resources, some of these sites were virtually destroyed from the point of view of any further value to mankind. Despite some noteworthy efforts to bring this process under rational control, it still goes on in a destructive way in many places. Because of the high industrial value of the raw materials obtained, mining has received a preference over virtually all other forms of land use, and the presence of high-value mineral resources has been used to justify the destruction of human communities, of living resources and of man himself – sacrificed again and again in mines where safety values were ignored and human labour was considered an expendable item to be valued only in terms of its immediate replacement cost.

Garret Hardin has pointed out how the science of economics has been used in explanation of the destructive consequences of industrialization. The deceptive word is 'externalities'. The costs of an industry that are not borne by the owners of that industry are known as the external costs or 'externalities' of that industrial process. In Hardin's words:

Reviewing the development of Western civilization during the past few centuries, we can see an interesting historical trend. To illustrate this let us list some of the more important identifiable costs of a hypothetical manufacturing operation, as an ecologist would view them.

1. Raw materials.
2. Labour.
3. Cost of raising and educating a labour force.
4. Cost of industrial accidents.
5. Cost of industrial diseases.
6. Cost of clearing up pollution of the environment.
7. Cost of preventing pollution of the environment.

From an environmental point of view each one of these costs is a legitimate charge against the cost of producing any particular product. From a historical viewpoint, however, the owners of business or industry have sought to transfer each of these costs to the environment and to the public at large. Socialist enterprise has differed little from capitalist enterprise in this process.

Far back in history, raw materials were obtained from the public domain – dug out of the earth, gathered from the fields, or cut from the forests. The manufacturer was not charged for them. This attitude still prevails in relation to some resources – for example, the pilfering of beach sand from the public beaches of the West Indies for use in the manufacture of building materials, or the removal of rock and gravel from public lands by private users. Eventually, where the raw material was in short supply relative to demand, the manufacturer sought by some means or other to achieve control over it. The institution of private ownership over resources was one such means.

The cost of labour was also externalized as far as possible. Slavery was initially a potential source of workers for the mine or factory – or for use in road construction, rowing galleys, towing barges, etc. But slaves must be fed and cared for or they cease to function, and therefore labour costs could not entirely become 'externalities'. Eventually it seemed more practical to hire labourers than to count on the uncertain yield of the slave market, but again the costs of breeding, rearing and educating these workers was transferred to a maximum degree to the community at large. The industrialist paid the minimum wage to keep the worker alive during the time he was working and assumed no responsibility for his childhood, old age, sickness or health, nor for his progeny.

We have only recently passed through the long struggle between labour and management, marked in some countries by socialist revolution, that led to internalizing the costs of the health and welfare of the worker, to an assumption on the part of industry of a concern for the long-term well-being of those who work in the production of its goods.

We are finally faced with the terminal phases of the struggle, the effort to have industry assume responsibility for the environmental consequences of its activity, and the owners or managers of industry resist violently the effort to internalize the 'externalities' of pollution control, of repairs for damage to the environment, or of the necessary steps to prevent damage to the environment. Socialist and capitalist industry alike seek to keep their production ledgers tidy by transferring to the total society the costs of preventing or repairing environmental deterioration. Yet, viewed objectively, these are surely costs of production.

It is not possible in any brief account to review the full extent of damage to the environment caused by industrialization. By far the greatest array falls under the category of pollution and amounts to the most serious environmental problem of the twentieth century – indeed we have reason to wonder if there can be a twenty-first century without its solution.

Pollution today differs from pollution in any earlier period because of the nature of the pollutants. This is not to say that the problems related to the earlier forms of pollution have been solved. They have not. The disposal of human sewage alone constitutes a major concern for all of the cities and villages of the developing world. The failure to develop effective systems for its disposal leads to the high rates of disease and mortality in many areas. Even in the most advanced countries, new pressures placed upon previously adequate disposal systems have created a new range of problems. Lake Geneva in Switzerland – once a clear, cold mountain lake – is clear no longer and has become a menace to the health of those who would make use of it; this is because of failure to find satisfactory ways for disposing of the human wastes from the growing populations of the cities and towns that surround the lake. The answer to the problem is known. The issue involved is the willingness of the communities concerned to bear the cost of the necessary sewage treatment.

But over the past half-century, and most markedly over the past three decades, each year has brought some new kind of pollutant, and our ability to cope with each of these has been tested. The turning point was unquestionably the Second World War. With the war's end, two major sources of pollution had been discovered that were virtually not in existence beforehand – the use of radioactive materials and the use of synthetic organic pesticides. Radioactive fall-out from the testing of atomic and hydrogen bombs, and later radioactive wastes from the development and use of radioactive minerals as a source of industrial materials and electric power, presented mankind with a new sort of pollution problem – one that could not be detected by ordinary human senses, yet was potentially fatal to all life. More than any other category of pollutant, we have with time learnt to control the output of radioactive materials into the environment. But in doing so, humanity has learnt a lesson. A pollutant does not have to be visible, to smell bad or to have immediate effects in order to be dangerous. This lesson proved valuable in dealing with the second category of Second World War 'externalities', the environmental effects of the new chemical pesticides. As described earlier, the discovery of the consequences of these chemicals took time, and the process brought the science of ecology into the public arena for the first time. It has so far been difficult to force the users of these materials to bear the full burden of their 'external' costs. How does one charge the farmers of California's San Joaquin valley, or the chemical industry that supplies them, with the wealth lost through exterminating the last of the California brown pelicans, or the last of its cormorants, murres or egrets? How much is a pelican worth? How much is the last pelican on Earth worth? We know their value is high. We suspect that in the long run it may be higher than we can today imagine. But how do we weigh it against so many pounds of cotton – even unwanted cotton?

Today we are overwhelmed with the array of dangerous pollutants being released into the human environment. In the United States, all the swordfish catch from the Pacific and a considerable proportion of the tuna catch has been condemned because of its high mercury content. What does this signify for the marine life of the world? Mercury, a heavy metal used for many industrial processes and purposes, is in its organic, methyl form toxic to all life if present in quantity. What is an acceptable quantity, a normal quantity? We have, as yet, little way of knowing. The waters around Japan, Sweden and the Netherlands are conspicuous for their high mercury content and, in Japan in particular, people have died because mercury has been released into rivers or estuaries. Is this a reasonable 'external' cost for the mercury-using industry? What proportion of the cost should be borne by society as a whole? How many people should be permitted to sicken, go mad or die? These are hard and real questions associated with industry today.

In the late 1960s, tens of thousands of sea birds died in the Irish Sea from causes that were unknown. Investigation showed that they contained a high percentage of those chemicals known as PCBs, polychlorinated biphenyls – close relatives of the chlorinated hydrocarbons that form the base of persistent pesticides. But PCBs are used in or are by-products of a wide range of industrial processes and appear in manufactured form in everything from cosmetics to herbicides. If they are indeed potentially harmful to man and his environment, how do we go about unravelling our complex industrial processes to provide for their elimination or control?

Perhaps the most serious question of all is, How do we prevent responsible industries, those that are willing to internalize the external costs – that are willing to accept the cost of pollution prevention and control – from being forced out of the market by irresponsible industries that assume no such costs and can in consequence market their products more cheaply? What do we do when industries are located in different nations, some of which, on the justification of economic necessity, refuse to impose any restrictive regulations on these industries? These are questions to which major world conferences are being devoted. We have yet to see them answered in a way that leads to confidence for man's future on Earth.

Urbanization

Cities have been on Earth for more than 5000 years and, during all of this time, there have been various types of environmental problems associated with urbanization, with urban growth, and the tendency for the inhabitants of rural areas to drift to cities in an effort to improve their material well-being. There has been a continual tension between city and countryside since the former is entirely dependent on an inflow of food and other essential materials from rural lands but at the same time, as a centre for manufacturing and cultural innovation, it has returned to the rural lands goods and ideas that were of value. Yet during most of human history, and up until very recent times, cities were small in size and most people had very little to do with them. Even the most famous cities of ancient and medieval times – Athens, Florence, Venice, seldom exceeded 50 000 in population. Some of the old cities of China were larger, and it was believed that Teotihuacan in Mexico may have held 100 000 people. Rome was an exception – there were few other cities of a million.

During the past century and particularly in the past few decades urbanization and its consequences have developed into major problems for all nations. In the developed nations most people are urbanized. In developing nations the former trickle of people from farm to city has grown to a flood.

Japan represents a recent trend in a nation that is now one of the most technologically advanced. In 1920 only 18 per cent of Japan's population was urban. By 1940 this had grown to nearly 40 per cent. Today approximately 70 per cent of the Japanese live in urban areas. During all this time, because of greater efficiency in production, farm output has increased even though far fewer people are involved in agricultural work. The city of Tokyo was established only during the twelfth century as the fortified seat of a warlord. In the eighteenth century Tokyo had become the world's largest city, with a population of nearly one and a half million. Its nearest rival, London, at that time had a population of 900 000. However, this was only the beginning of Tokyo's expansion. In 1920 there were more than three million people in Tokyo. By the start of the Second World War, the population had grown to over seven million. Today the city itself has over ten million and the metropolitan region over fourteen million. Depending on where the metropolitan boundaries are drawn, either Tokyo or New York has the title of the world's largest metropolitan area. The growth of Tokyo continued despite the widespread devastation wrought by the fire and earthquake of 1923 and the fire-bombing of 1945. Similar growth has been demonstrated in the Osaka–Kobe metropolis as well as in other, smaller cities.

In the most urbanized nations such as the United Kingdom and Australia more than 80 per cent of the people are urban. This appears to represent an upper limit of urbanization under present technological conditions and seems unlikely to be greatly surpassed. Urbanization has involved a tendency for individual cities not only to grow in size but also to flow together to merge into greater urban complexes for which the term megalopolis has been devised. Such a megalopolis is taking shape in eastern North America in the region between Boston and Washington. Constantin Doxiadis has recognized thirteen of these megalopolitan areas in the world:

1. Japan – Tokyo, Yokahama, Osaka, Kobe
2. China – Shanghai, Nanking
3. China – Peking, Tientsin
4. China – Shenyang, Dairen
5. China – Hong Kong, Canton
6. Indonesia – Djakarta, Bandoeng
7. Egypt – Cairo, Alexandria
8. Eastern United States – Boston, Washington
9. Great Lakes – Detroit, Chicago
10. California – Los Angeles, San Diego
11. Rhine – Amsterdam, Hague to Cologne, Dortmund
12. England – London, Manchester
13. Italy – Milan, Turin

One may argue with this classification and may wish to either add or subtract areas from Doxiadis's list. Nevertheless, the tendency to expand and merge may be observed in most metropolitan regions to a degree where it is now difficult to separate formerly distinct, and indeed unique, urban centres. There are now at least twenty-five major metropolitan areas with populations in excess of three million people each. No doubt the latest census will reveal more.

The possibility of restricting and controlling urban growth has exercised the minds of rulers and governments over many centuries. In this connection it is worth noting that it became official policy in the Soviet Union in the 1930s to restrict the growth of large cities. In the development of the General Plan of Reconstruction for the City of Moscow in 1935, limitation of urban growth was a planning objective. The target population for greater Moscow was set at 5 000 000 and this was to be achieved by natural increase from the 1935 level of 3 660 000. Net immigration was to be eliminated. Despite this plan and all efforts to make it work, the population in Moscow proper, inside the green belt, had grown to over 6 000 000 in 1960. In addition, a row of *sputniki*, satellite cities outside the green belt which were functionally a part of Moscow, had an additional population in 1960 of over 2 000 000. Within the green belt proper lived nearly 1 000 000 more people, making a total for the metropolitan region of over 9 000 000. With this in mind, one may wonder at the prospects for limiting growth of other major cities in countries where governments have little power to control land use or movement of peoples.

There is little doubt that the sudden surge of urbanization during the past few decades has had many aspects of an environmental tragedy. The situation has been summarized in a statement prepared for the Biosphere Conference by Dr Guy Gresford of the United Nations.

According to his information, the growth of populations in recent decades has been accompanied by the spread of urbanization. Forty per cent of the world's people now live in urban areas. In somewhat more than half a century, if present trends continue, urbanization may well have reached its maximum and the great majority of people will live in towns and cities. The rate of urbanization, however, is more rapid in the developing nations. According to national estimates, in 1920 the urban population was a hundred million in these countries. By the year 2000 it may well have increased twentyfold. In the developed nations, by contrast, the urban population in the same period will have increased fourfold. These are projections which for many reasons may not be realized, but they illustrate the size of the problem.

Urbanization is not in principle destructive to the environment. With proper planning and control, and if it were proceeding at a slower rate, it could enhance and not detract from environmental quality – by relieving pressure on rural lands, by providing goods and services in quantity and diversity, and by providing new and attractive habitats and ways of life. However, in most areas, governments have neither prepared for, nor have they been able to cope with, the mass migration into urban areas. In the large cities, slums of the most wretched nature have often become the environment of people who once lived in greater dignity and better health on rural lands. Pollution of air, water and land, concentrated in urban areas, have become universal problems, threatening man's health. Diseases associated with urban living in developing nations have increased greatly despite advances in medicine. And, finally, the noise and congestion of cities adds to physical and mental distress.

The environmental impact of urbanization is of two major

kinds, that external to the city proper and that within the city. Externally the greatest effect of spreading urbanization has been the intensification of pollution which spreads from the cities outward to have its effects throughout the biosphere. Secondly, the spread of cities and the transportation networks that connect them have affected all of the lands that surround the metropolis and those through which transport corridors pass. Because urban uses are given a higher economic value than other, less intensive forms of land use, they tend to displace these other uses. Lands that could be best used for agriculture, forestry or recreation are often used for urban purposes because of their accessibility or ease of development. Other lands that could better be used for urban development may be by-passed in the process. Lack of planning and control over land use are the reasons for this contradiction but, in the face of extremely rapid urban growth, planning and control become difficult.

The conflict between urban uses and high, non-urban values has been particularly severe in those cities that develop along the edge of lakes, rivers or estuaries. Often the physical spread of the city leads to dredging and filling in of previously valuable and productive aquatic areas. Always the consequences of urban pollution on such environments is severe. Along sea coasts and estuaries in particular, damage to the aquatic environment can have far-reaching consequences. Often the productivity of major ocean fisheries and of broad areas of the open ocean depends upon the continued functioning of the estuarine and near-shore region. When these are damaged by urban growth or pollution, major natural values are affected.

The modern metropolis draws, for its resources, upon great areas of land and water distributed widely throughout the biosphere. Thus, for its water supply alone the city of Los Angeles draws on watersheds hundreds of miles away. Its other needs, for food and fuel, for example, are supplied from around the world. In turn the products and by-products of urban life ramify throughout the biosphere. A balanced relationship between the city and its global environment is therefore of major importance for any programme of rational use and conservation of the total human environment.

Although the external effects of urbanization are impressive most people are directly affected by the environment within the city itself. Although this has rarely been of high quality, it has under the pressure of too rapid growth and the consequent breakdown in urban functions become increasingly unsatisfactory to the people involved. Pollution has, of course, been a major contributer to the decline of environmental quality within the city, but it is not alone. Crowding and congestion in themselves can have wide-ranging physical and psychological consequences which we are only beginning to understand. Housing has generally been inadequate in most rapidly growing cities. Not only has there been a spread of slums within the city proper, but most cities in the developing world have developed a ring of shanty towns in which even the most primitive urban services are lacking.

The problems of urbanization are now far beyond the capacity of city governments to handle. They have become national problems which require a high degree of international cooperation if they are to be successfully surmounted. The cost of providing even the most minimum, essential environment that will permit a healthy, productive life for the city dwellers of the world must be measured in many hundred thousands of millions of pounds. No other environmental problem, save the related one of pollution control, will require such a major share of effort and energy in the decades that lie ahead.

Major engineering works

In ancient Egypt certain of the kings or pharaohs decided to achieve a level of immortality denied to ordinary human beings. Mustering their subjects in their tens of thousands, they set them to work on great monuments that were to endure over the millenia and be admired by generations of men to come as engineering masterpieces. There is no evidence that construction of the pyramids accomplished any serious degree of environmental damage. There is little doubt that they kept people employed and thus may have prevented some turmoil and strife within the Egyptian domain. On balance they added some small increment to the quality of life in Egypt for those generations that were to follow the builders.

Designers, architects and engineers were similarly employed

throughout all of ancient times. In Java and Cambodia, Greece and Rome, Mexico and Guatemala great monuments to the glory of God or the might of the emperor grew from the ground to win the admiration of the people. Today, many long centuries after, when even the reason for their construction may have been forgotten, tourists come by the millions to pay their homage to the Acropolis and the Colosseum, to the Temple of the Sun and the ruins of Angkor, to Macchu Pichu and Borabadur. But not all of man's great works of engineering have been so benign nor have their contributions to the environment been so universally favoured.

Up until the past century the ability of man to accomplish major changes to the face of the Earth was limited. Admittedly, great engineering feats were accomplished, not the least being the construction starting in 214 BC of the Great Wall of China. But such efforts proceeded slowly, there was always time to reconsider, change and redress one's mistakes. Perhaps the last of the major engineering works to be constructed primarily by hand labour with the assistance of domestic animals was the Suez Canal which in 1869 linked together for the first time in recent geological history the Red Sea and the Mediterranean. This was also an engineering work that was to have biological consequences of a long-lasting nature that were not, and could not conceivably have been, given any attention at the time the canal was constructed. Primarily these involved a slow shifting of the Red Sea fauna into the Mediterranean with a potential consequence in the displacement of Mediterranean species. The effect is a relatively recent one and is related to the decreasing salinity of the Bitter Lakes which stand between the Mediterranean and the Red Sea ends of the canal. Originally these provided a massive saline barrier to the movement of species from one sea to another. With the passage of time and the flow of water through the canal the effectiveness of this barrier has gradually been removed. It is as yet too early to tell what the long-range consequences of this faunal shift will be.

Today we have the capacity to make sudden, massive changes to the face of the Earth. We can and do move mountains, fill valleys, shift rivers from one region to another, drain old lakes and fill in new ones. We do these things quickly and often, indeed usually, without adequate study of the potential consequences.

One object lesson that shows both the potential consequences that may develop in the Mediterranean as the Bitter Lakes barrier disappears, and also potential consequences of other proposed canals is to be found in the North American Great Lakes. Lake Ontario in the Great Lakes system has long been connected to the oceans through the Saint Lawrence River. Those species of fish that regularly migrate from salt water to fresh have long had access to Lake Ontario; among them is a parasitic fish of a primitive form, the sea lamprey. The upper Great Lakes, however, were isolated from all sea-run species by the Niagara Falls in the river connecting Lake Erie with Lake Ontario. But the development of the Great Lakes as a major industrial region for the United States and Canada led to a desire to provide access for ocean-going ships to all of the Great Lakes. Niagara Falls was one major barrier that prevented this. To by-pass Niagara Falls, the Welland Canal was constructed in the nineteenth century. This allowed ships to move from the 'Saint Lawrence system on to the upper Great Lakes. Unfortunately, it also provided access to the upper lakes to the fish fauna from Lake Ontario. One of the first to take advantage of this was the sea lamprey.

Fish species in the upper Great Lakes had developed in isolation from lampreys and were unable to adapt to their presence. The lamprey preyed particularly upon lake trout and the whitefish, both of major commercial importance and the basis for important fisheries. The Welland Canal was completed in 1829, and the lamprey entered Lake Erie. It next navigated the Saint Clair River and entered Lakes Huron and Michigan. Finally, in the 1940s it was spread throughout the entire Great Lakes system. The lake trout and whitefish populations were decimated and a fishery valued at twelve million dollars was virtually eliminated. Production in Lake Huron and Michigan alone fell from 3900 tonnes to less than 12 tonnes of lake trout as a result of lamprey predation alone. Since the 1940s a multi-million-dollar lamprey control programme has been launched, involving fisheries biologists of the United States and Canada. After years of work a chemical poisonous to lampreys but harmless to other fish has been discovered. It appears that by attention to the small tributary streams in which the lamprey spawns, the problem may now be eliminated. But tens of millions of dollars have been lost unnecessarily. With adequate foresight, the Welland Canal could have been constructed with proper barriers to lamprey movement. Admittedly, in 1829, such a problem could not have been foreseen. In the 1970s, such problems can be anticipated.

It is now proposed to construct a sea-level canal through Panama connecting the Pacific and the Caribbean directly. The means are available to do such a job quickly. Such a canal will differ from the existing Panama Canal in that it will lack the series of locks by which ships are lifted over the Panama isthmus in moving from one ocean to another. More important, it would lack the major freshwater barrier which the Gatun Lake now interposes between the Pacific side of the canal and the Caribbean. As the canal is planned at the present time, no barrier to the movement of fish or other forms of aquatic life from one ocean system to the other would be provided. Yet the Pacific and Caribbean fish fauna have developed in complete isolation from each other during many hundreds of thousands of years. It is impossible to predict what will happen if they are connected, but it is probable that adverse consequences, particularly upon the more isolated Caribbean biota, will result. Biologists have strongly recommended that the canal must not be built unless appropriate barriers to inter-oceanic movement of marine life are provided. It remains to be seen, if their advice will be followed.

Starting in the 1930s, work was initiated to construct a major barge canal through the state of Florida, from Yankeetown on the Gulf of Mexico through to the Saint Johns River on the Atlantic Coast. The chief justification for the canal was the provision of cheaper transport for bulk raw materials from the Gulf Coast ports of the United States to the Atlantic ports, and also cheap transport for raw materials such as phosphate rock mined locally in Florida. Work on the canal was sporadic, and it was not until the 1960s that funds were provided in sufficient quantity to allow for a serious effort to be undertaken. At that time environmental scientists became alarmed over the potential consequences of constructing the canal. The proposed route would lead the canal through one of Florida's few remaining wild areas and one of its most scenic rivers, the Oklawaha. It would cut into the Florida aquifer, the source of fresh water for many communities in central and southern Florida. Further investigations showed that some degree of pollution of the canal waters and its major impoundments would be unavoidable, and this would be led into the aquifer. Furthermore, the virtual impossibility of preventing invasion of the canal by the various water weeds that were already well established in other Florida water bodies was pointed out. These and other arguments were listened to, but apparently not considered, by those charged with digging the canal, nor by the canal's political supporters. It was not until the creation of the Council on Environmental Quality in the office of the President of the United States in 1969 that the arguments against the canal were fully evaluated. When this happened, the President ordered work on the canal to halt, but already fifty million dollars had been spent and essentially wasted, on this potentially disastrous engineering scheme.

Most of the major engineering plans that have important environmental consequences involve water and its many potential uses. The various problems associated with dam development on major rivers have been discussed earlier in this volume (Chapter 1). Such problems have developed in relation to the High Aswan Dam in Egypt. Related problems have developed in relation to most of the newer high dams in Africa. Studies of the proposed Mekong River developments in South-East Asia suggest that the adverse consequences, both social and ecological, may outweigh the benefits expected from this multi-million-dollar effort. Certain proposals to reverse the flow of northward-moving rivers in Siberia and Canada can have consequences of such great magnitude as to deserve full investigation by environmental scientists before any major engineering work is undertaken. The same can be said for major impoundments suggested for the Amazon basin in South America.

During recent decades, those concerned with the financing and construction of major engineering developments have listened to the advice, primarily, of economists and engineers. Rarely are ecologists or other environmental scientists consulted during the planning stages. More rarely is their advice followed. Consequently a number of serious environmental problems have developed which could have been avoided. The danger now is, with our newly developed ability to accomplish engineering miracles, we may trigger a series of environmental catastrophes. There is no need for this to occur.

Intensification of land use

The greatest changes in the biosphere during the past half-century have come about, not as a result of the processes of urbanization, industrialization or engineering as such, but rather because of the intensified use that is made of productive lands. The intensification of rural production complements the increase in urbanization and industrialization. It would be virtually impossible for one to proceed without the other. It has been pointed out that industrialization became possible in eighteenth-century England because of the high efficiency and increased productivity of English agriculture at that time. This freed people from the necessity of earning a subsistence from the land and allowed them to seek other employment. It provided the food to maintain the workers in the mines, mills and factories. A country without an agricultural surplus could not have made such a step forward. Similarly, the increase in agricultural efficiency, particularly since the Second World War, has touched off the new wave of urbanization and technological advance. The increase in the percentage of urban dwellers is accompanied, or rather preceded, by a decrease in the number of farmers. It no longer takes many hands to grow enough sustenance for all. Intensification has taken place in virtually all productive uses of the biosphere – from the growing of trees to the capture of whales. Three categories of human activity will be examined here: forestry, livestock management and agriculture.

Forest production

It was noted in the preceding chapter that, from an early time, man has seemed an enemy of forests. With the new power and the demands of industrial civilization, forests tended to disappear, to be replaced by other forms of land use. For most of man's time on Earth, forests were regarded as a natural resource to be mined in much the same way that one takes gold or coal from the Earth. The idea that forests might be managed for sustained production was slow in coming. To a considerable degree it was brought about by economic necessity at a time in European history when a shortage of industrial wood was first experienced.

The idea of forest conservation and sustained production appears to have arisen simultaneously in France and Great Britain in the seventeenth century. In 1664, John Evelyn, the founder of the Royal Society of London, published a book entitled *Silva: or a discourse on forest trees*. He noted the disappearance of the woods in England and proposed the idea of forestry as a science. At approximately the same time in France, Colbert, minister for Louis XIV, produced the *French Forest Ordinance of 1669*, in which the need for halting the disappearance of forests in France and the need for a sound programme of land management and forest conservation were advanced. The remaining woodlands of Europe were to be brought increasingly under management, forestry was to be established as a profession, and the foresters of the Americas and other lands received their first training at European forestry schools. Gifford Pinchot, for example, who did so much to advance forestry in the United States, received his education at the French forestry school at Nancy.

The new concept of forest utilization and management was based on the belief that forests were a renewable resource; that if properly cared for, forest trees would replace themselves as they were cut over. With care, there could always be a new crop of trees growing up in place of those removed, and the yield from a well-managed forest could be permanently sustained. In some parts of the world, notably North America, natural reproduction from forest trees was encouraged. In Europe a tendency developed toward replanting cutover lands with seedlings grown in forest nurseries. This led, inevitably, to a tendency to plant the most favoured tree species, rather than the mixed variety that grew in a natural forest. In Germany in the nineteenth century this led to a forest monoculture; areas were planted, for example, with successive generations of the desirable spruce trees. Under such practices, however, several problems developed – those revolving around forest pests which concentrated on the single species that they too preferred, and difficulties with maintaining soil structure and fertility. Succeeding generations of monocultural spruce forests were found to give lower yields, and German forestry in the 1930s and 1940s shifted back towards an emphasis on the natural forest, with its full range of species, in place of the artificial plantation.

However, in areas of the world where natural forests could not provide the fast-growing softwood species that are in high demand for industry, forest plantations were to be emphasized. In Australia, New Zealand, South Africa and, later, throughout the tropical world, forests of rapidly growing exotic conifers and eucalypts were planted in place of native species. These provided high initial yields of valuable timber, higher than could be expected from natural forests. Whether or not the yields can be sustained over the long run has yet to be fully tested. However, forest science has proceeded a long way, and the means for fertilizing soils and other procedures for maintaining soil structure are now well understood. It seems unlikely that the German experience will be widely repeated. The new plantation forests further serve to remove pressure from natural forests, which can then be managed to suit a wide variety of human requirements, rather than the strong emphasis upon wood production that would otherwise be exerted. But when the new plantations spread too widely at the expense of natural woodlands, the results can be deplorable from any point of view other than that of timber production. Plantation forests lack diversity in a faunal as well as a floral sense and are the equivalent of farming crop lands spreading into areas that would otherwise be maintained in wild vegetation and animal life.

Today, through the application of new technology to forest lands, it is possible to grow sustained crops of trees in far greater volume than would formerly have been thought possible. The trees now yield a variety of products not previously provided – pulp, paper and plastic production, as well as the production of numerous varieties of synthetic boards, has surpassed in importance the yield of lumber and construction timbers. Furthermore, the new products may be obtained from a wide variety of tree species so that the full timber production of a forested area may now be utilized, whereas previously only a few favoured species would be cut. Where properly controlled and managed, such broad utilization of forests is a means for obtaining long-term production to meet human needs. Where uncontrolled, it becomes a particularly savage form of land destruction. Unfortunately in the tropics the old attitudes toward forests still prevail. Great areas of valuable forest are still cleared for short-term gains in food production or forestry. Other areas are 'mined' for a few high-value trees with no thought for permanent production. The techniques for successful management of the complex and diverse forests of the tropics have yet to be fully devised.

Livestock management

The history of exploitation of rangelands and range forage through the use of domestic animals has been depressing indeed. Although primitive nomads may have lived in some balance with their environment, the more modern commercialized exploitation of rangelands initially sought no such accommodation. Just as in the exploitation of forests in areas where conservation and management has not been understood, so in the use of rangelands man took what he needed, while nature was expected to look after herself. The story of the exploitation of the drier rangelands of Australia is illustrative.

Australia was colonized by the English in the eighteenth century, and the first settlers brought domestic grazing animals with them. Within two decades after settlement there were 25 000 sheep in Australia, mostly concentrated near the coast of New South Wales. By the early nineteenth century the original pastures were almost entirely devastated by overgrazing, and the pastoral industry moved to the interior. Seventy years after settlement there were 20 000 000 sheep in Australia. By the end of the first century after settlement, their numbers had reached 106 000 000 and they were spread throughout all available rangelands. Where sheep went, overgrazing followed and the ranges began to collapse. When droughts came there was no forage reserve and die-offs began. Between 1891 and 1902, 53 000 000 sheep were lost, not counting the lambs that might have been produced. The total stock in Australia was halved.

In one year of drought alone, 18 000 000 sheep died. Again, between 1911 and 1915, the number of sheep on Australia's ranges decreased by 25 000 000, with 10 000 000 dying in one year. In 1919, a further 11 500 000 sheep died. In the process the rangelands of Australia were drastically modified. Many have not recovered their original carrying capacity. Nevertheless, the Australian sheep industry has since thrived. Much of the gain has come through careful management and restoration of the grasslands in the better-watered areas of the country. Much has come through more intensive care of the animals themselves. It has been learnt that far more money can be gained by producing high-quality forage on good land and feeding it to high-quality livestock, than by turning hardy animals loose to plunder and destroy natural rangelands. Admittedly, the lesson has not been learnt too well. Much of the drier rangeland of Australia is still poorly managed, and some destruction done in the past is virtually permanent in its effects. But basically, the pastoral industry, like the forest industry has moved from 'mining' to 'cropping'. Pastures and animals are conservatively managed for a sustained and permanent yield.

The Australian experience can be duplicated in California. By the middle 1930s it was believed that the carrying capacity of California's rangelands had been reduced by one half, and a similar estimate was made for all of the rangelands of the western United States. Nevertheless, livestock production from California today is higher than during the period of exploitation in the past. Carefully managed pastures, the integration of livestock with crop farming and better breeds of livestock given greater care all contribute to high yields per hectare. It has paid more to manage than to plunder.

Agricultural production

The shift from an extensive, subsistence, low-yield agriculture to an intensive form of modern agriculture has produced more striking results in the world than the changes brought about in forestry or livestock management. Nevertheless, the tendency has been the same – concentration of investment upon the better-quality lands or the most productive varieties of domesticated species. Characteristic of modern agriculture has been mechanization – the use of tractors and a variety of farm machinery in place of human labour or domestic animals. It follows that such mechanization is based upon the availability of fuel in quantity and is more suited to large farming units than the small farm. Heavy use of fertilizers and other agricultural chemicals, including pesticides and herbicides, has characterized agriculture, particularly since the Second World War, together with a strong reliance upon improved varieties of crop plants specially bred for high yields, response to fertilizer and disease resistance. Widespread use of irrigation, even in areas of moderate rainfall, is also associated with modern agricultural practice. All of these changes have brought a decreased reliance on natural soil fertility and a tendency to favour agricultural soils which are easily worked by farm machinery, hold up well structurally under cultivation, have a good ability to absorb water and fertilizers and release these to growing plants. Relatively high capital investments are required, but these bring high responses in yields and profits. Nevertheless, the large landholder has a strong competitive advantage over the small farmer, since costs per hectare decline with increasing size of the farming unit.

The appearance of new, high-yielding hybrid cereals during the late 1960s and early 1970s, particularly the new rice varieties developed by the International Rice Institute in the Philippines and the new wheats developed in Mexico by the Rockefeller Foundation, have brought about what has been termed a 'green revolution' – an enormous increase in yield of cereal grains. The extent of these agricultural gains has been of such a magnitude as to cause optimistic statements on the permanent removal of the threat of famine from the world and, in 1970, led to the awarding of the Nobel prize to N. E. Borlaug, one of the leaders in the development of hybrid grains. However, as Borlaug has stated, the new grains only serve to buy a little longer time during which population growth must be brought under control. Without such control the threat of famine will always remain on the horizon.

The changes in agriculture that have transpired have so far worked primarily for the benefit of the technologically advanced nations, which were best able to mobilize the machines, the chemicals and the trained manpower. In these countries food surpluses have replaced former scarcities, and it has been possible to cut back on the agricultural acreage to a marked degree while greatly increasing agricultural output. Furthermore, the increased efficiency of agriculture has not only permitted, but has almost forced, an increase in urbanization. Relatively few people are needed for work on the modern farm and a rural population surplus was forced to seek work in urban areas. In the technologically advanced countries it has mostly been possible to absorb displaced rural workers in the urban–industrial framework, but difficulties have been encountered with those who were poorly educated and lacked skills. As the new agricultural advances have spread to the non-industrialized world, they have created new problems. Displacement of agricultural labour and of the smaller or less efficient farmers has also transpired, but no place has been available for them elsewhere in the national economy. Lacking the funds to purchase the newly abundant foods or to otherwise take advantage of the benefits of technology, they have either drifted into a life of poverty in urban slums, or have been forced into a minimal subsistence existence on marginal farming lands.

The other environmental consequences of the new agriculture have not been entirely benign. Although in some countries the new practices have released, for other purposes, land that was formerly needed for crop production, this effect has not been universal and, for the reasons noted above, has scarcely transpired at all in the developing countries. The use of pesticides and herbicides has brought consequences that we have already examined. The full magnitude of the environmental effects of biocides seems almost certain to prove to be much greater than we can thoroughly substantiate at present, and there is little reason for optimism. Heavy use, or, more accurately, misuse, of nitrates and phosphates as fertilizers has brought a new range of problems, particularly involving the excessive enrichment or eutrophication of streams and lakes. The increased use of irrigation in tropical areas has brought a spread of waterborne diseases that cannot yet be controlled. All of these environmental effects could be, and indeed many were, predicted by ecologists before the event – but in the rush to solve the immediate problem of food supply and to gain economic advantage, such predictions were generally ignored. It is now necessary to take them into account before damage becomes more severe. This does not mean giving up the benefits of modern agriculture, but of modifying the practices to control or eliminate the undesirable effects.

One of the more insidious effects of modern agricultural practice has been the loss of genetic diversity among domesticated plants and animals and their wild progenitors. The tendency to concentrate upon high-yield hybrid cereals has often caused the loss of the wide range of cereal varieties that existed in the country before the hybrid appeared. These are essential in maintaining the genetic variety from which the hybrids of the future will be developed. In the absence of such variety, the new crops, no matter how successful initially, remain very vulnerable to attacks from various diseases or insect pests. Disease-resistant varieties keep their resistance only until the time when the disease organisms produce in turn their new varieties able to overcome the physical or biochemical defences of the new crop plant. This usually occurs within a few years after a new variety is introduced. Thus there is a continual need to produce new strains faster than pests catch up with the weaknesses of the old. In order to even compete in this race, let alone win it, it is essential to have a wide range of genetic material available. Yet this genetic material is quickly lost when the introduction of new hybrids is carried out without a corresponding effort to rescue and maintain the older crop plants.

Chapter 5
The Movement Towards Conservation

Damage to the human environment brings its own reaction. Not only is there a feedback in a physical sense, which leads to an impairment in productivity or a decline in the quality of life, there is also what may be more important, a psychological reaction against the processes that cause damage. In earlier times, when environmental damage was limited, the reaction against it was limited but, in modern times, when damage has become increasingly obvious, severe and inescapable, the reaction has become strong and widespread. This reaction against those processes that cause deterioration has resulted in the movement toward conservation.

If man ever lived in a Garden of Eden, we can be quite certain that there were no conservationists there. In those areas where man has lived for a long time in a reasonable balance with nature, conservation has not been recognized as a distinct social movement, and its few adherents were to be found among those with sufficient training and insight to recognize problems of which the general populace was unaware. It could be said that in such areas conservation was a normal component of existence – good land use and care of the environment were taken for granted. It was in those places where damage to the environment was sudden and far reaching, and therefore obvious to a wide segment of the population, that the modern conservation movement had its origins. Such a place was North America in the latter part of the nineteenth century. Man's impact on the American environment was devastating at a time when awareness of such changes on the part of the public had been made possible through the development of improved systems of transport and communication. The impact of the sudden destruction of the great herds of buffalo that had roamed the American plains, of the devastation of forests that once had seemed endless by log-

ing and fire, of the droughts and dust storms that raged over
once-verdant fields came at a time when the people were first
becoming aware of the existence and value of these resources.
As a reaction, there was widespread political support for any
move that would protect and restore the environment. The word
conservation was first used in its modern sense, and government
agencies charged with the task of protecting and managing
natural resources were given broad power and responsibility.

By contrast, the concept of conservation has been slow in
becoming accepted in western Europe. There the process of
environmental change was gradual. Man had achieved stable
balances with the lands and resources that supported him and
there seemed little need for basic changes in long-established
patterns of land use. Admittedly, those who were concerned
with the preservation of wild nature were disturbed when animal
species became extinct or biotic communities disappeared, but
such changes came slowly and caused little public reaction.
Only in the past few decades, when the impact of new technologies has disrupted the old patterns of European life, has there
been growing support for conservation in Europe, and this
concern is largely directed against the growing dangers from
pollution and, in some places, against the uncontrolled spread of
urbanization.

Outside North America, the lands in which conservation
became most quickly accepted were those undergoing a similar
process of rapid change. In Africa and Australia the effects of
modern technology upon lands and wildlife were as destructive
and rapid as they had been in the United States. The need to
halt the damage and begin the processes of repair and management for the environment was to become equally apparent.

Along with most other recent activities, the conservation
movement is based on ideas and activities that have their origin
far in the past. It has often been noted that Plato attributed a
decline in fertility in ancient Greece to the deforestation of the
mountain watersheds and the consequent soil erosion and decline
in the reliability of stream flow. Further, Plato was one of the
first to recognize the importance of the ecological idea of diversity, the principle of plenitude – that the world becomes better
as it contains more things and more variety. Edward Graham has

observed that the first wildlife sanctuary may well have been created by Sennacherib in an area near Nineveh. Such sanctuaries, usually set aside as hunting grounds for royalty, played a useful role in the protection of wild species and provided a traditional basis for the later establishment of completely protected parks and reserves as well as public hunting grounds.

The principles of agricultural conservation, the sound management of farming land, have their origin in trial-and-error practices dating back to the Neolithic. Some of these concepts appeared in written form in Egyptian papyri, and were later to be developed scientifically in the writings of such Romans as Cato, Columella, Pliny and Tacitus. As noted earlier, modern concepts of forest conservation and general land management were expounded in seventeenth-century England and France by Evelyn and Colbert. Further development of the ecological basis for conservation of land and biotic communities may be found in the writings of Buffon in France in the eighteenth century and Alexander von Humboldt in Germany in the early nineteenth century.

During the eighteenth century problems related to human population increase were examined in a series of papers and a controversial exchange of ideas between M. J. de Condorcet in France, William Godwin and Thomas Malthus in England. The Marquis de Condorcet and William Godwin were in many ways the philosophical forerunners of those who today appear to have unlimited faith in human institutions and technology as means for removing environmental limitations upon the growth and expansion of human populations. Against their ideas the famous *Essay on Population* of Thomas Malthus was directed. Populations, Malthus believed, must always increase faster than their means for subsistence can be increased. They will, in consequence, press upon the limits of their environment with results to be measured by the increase in vice and misery. Such increase could be prevented, Malthus was willing to concede, through the exercise of moral restraint, but for this he saw little hope. The gloomy outlook of Malthus found little support in the optimism and expansion of eighteenth-century Europe, but with the population problems of the twentieth century there has been a widespread revival of interest in his work.

The most striking characteristic of the conservation movement of the late nineteenth and twentieth centuries is the concern for wild nature. The gains made by human populations and technology have been wrought at the expense of the wilderness. Natural biotic communities, unmodified by man, have been restricted in extent by all of man's activities, but it was only with the recent surge of population growth and advanced technology that the future existence of such wild areas seemed to be seriously endangered. Wild animal life has suffered many setbacks at the hand of man, but only recently have widespread extinctions begun to occur. Admittedly, species became extinct long before man appeared on the scene, but such extinctions in the geological past were of a different nature. Other species always evolved to replace those that had been poorly suited to survive natural changes in the environment. Extinction today is extinction without replacement – a tendency toward a global impoverishment which was not present in the past. Consequently there has developed a widespread movement to protect and restore the natural environment. This has taken many forms – the establishment of natural parks and reserves, the institution of rational management of wildlife and wild vegetation, and studies of the means for restoring, restocking and replanting areas that were devastated in the past.

National parks

The idea of protecting natural areas of outstanding beauty, or of great interest, through the exclusion of all forms of land use likely to destroy or impair these values, had its first major implementation in the United States. This occurred at a time when the destruction of natural landscapes and of wildlife was proceeding at a rapid rate over much of the country, but also at a time when the availability of land was relatively great in relation to the foreseeable needs of the human population. In 1872, a group of far-sighted individuals were able to persuade the Congress of the United States to set aside an area of land surrounding the scenic wonders of the Yellowstone country of Wyoming and proclaim it as Yellowstone National Park. Such a park had a dual purpose – to preserve in an unimpaired form the scenery, wild vegetation and wildlife of the area, and to make it available for human recreation as a 'pleasuring ground for people'. Yellowstone was the first of what was to be a series of national parks, national monuments and similar protected areas to be established throughout the United States, providing protection for most of the more spectacular wild scenery and for distinctive and unique natural communities.

The national-park idea spread from the United States into other countries. Great areas of wild land inhabited by impressive populations of wildlife were to be given protection throughout the Americas, Africa and Australia, and eventually in most countries. In the Union of Soviet Socialist Republics, starting in 1924, a series of national reserves (*Zapovedniki*) equivalent to national parks were to be established. These include such impressive areas as the Pechora–Ilych Reserve of 721 322 hectares in the northern Urals and Pechora river area and the Zakataly Reserve of the high Caucasus Mountains of Azerbaijan, along with fifty other areas of national-park quality. In western Europe are such impressive parks as the Gran Paradiso of Italy, which had its beginnings in a royal hunting reserve which was changed to a national park in 1922. Starting in 1909, Sweden established the first two of its three largest national parks in Lapland. These nearly adjacent areas protect more than 500 000 hectares of Arctic flora and fauna.

Admittedly there is still a great gap between idea and practice in many countries. It is relatively easy for any government to proclaim a national park in a wild area in which few people have any practical interest. It is a far different thing to provide adequate protection and management at a time when the resources of a once-remote area begin to be in demand. Many parks exist on paper only, in documents on file at a national capital. In the field, nothing has happened and the resources of the 'national park' receive no more protection than those of any other land area. In the nations more advanced in this respect, a well-trained and dedicated body of people are employed by a public agency specifically charged with the protection and management of park areas. Elsewhere, national parks may be the concern of a government bureau that has neither the interest nor the ability to cope with this difficult task. Once it was safe to allow national parks to take care of themselves. Now, with increasing pressure upon all lands, and with great demands for areas for public recreation and tourism, the protection and care of a park is a difficult task requiring a wide variety of skills and expertise.

Scientific reserves

National parks have been, from the outset, areas made available for public outdoor recreation and use, provided that the forms of recreation and use were not destructive to the values that the park was created to maintain. Such public use, by its very presence and the need for facilities to accommodate it, creates various forms of disturbance in natural areas. The need for areas to be protected from all disturbance so that they may be used for the long-term study of wild species and ecological processes has long been apparent to scientists. In response to that need, many countries have designated special scientific reserves in natural areas to be used exclusively for scientific purposes. The term *strict nature reserve* has been applied to an area from which all forms of use are excluded except those forms of scientific investigation approved by the agency with jurisdiction over the reserve. Some of the larger national parks, such as *Parc National Albert* in the Congo–Kinshasa, have functioned in whole or part as strict nature reserves.

The need for protecting representative samples of the world's ecosystems in reserves set aside for scientific studies received greater emphasis with the institution of the International Biological Programme (IBP) in the middle 1960s. As part of this programme, governments have been encouraged to set aside such reserves, and an inventory and evaluation of the current world situation in relationship to the need for such reserves has been undertaken. A particular emphasis has been placed upon the need to set aside undisturbed islands throughout the oceans of the world and, in 1971, an international convention establishing a number of 'islands for science' was drafted and made available for consideration by governments.

Wilderness areas

Although many of the larger national parks in the world are maintained for the most part in a wilderness condition, there has been a strong feeling in some countries that additional areas need to be established and permanently maintained in a wilderness state, without roads or any other form of development. Unlike the strict nature reserve, such areas would be available for public use, but only by those willing to use primitive means of travel – hiking, travel by motorless boat, horseback riding, etc. – and to get along without any form of established camping grounds or recreational facilities.

Wilderness areas were established by administrative decree within the national forests of the United States during the late 1920s. A formal system of national wilderness areas was created by act of the United States Congress in 1964. As yet, however, this concept has not been formally adopted by other countries.

Wildlife conservation

The need for protecting wildlife has provided the motive for the establishment of many national parks and reserves throughout the world. Most of the national parks in Africa had their start as game reserves, areas intended to protect wildlife, but from which uses not considered deleterious to wildlife were not necessarily excluded. Only later was the more restrictive designation of national park adopted.

Many countries have various categories of wildlife refuges, sanctuaries and game reserves from which hunting is excluded, and on which various forms of management intended to protect or propagate wildlife may be undertaken. Some such reserves represent lands in government ownership set aside exclusively, or primarily, for wildlife protection. Other refuges are established on private land or land in use primarily for purposes other than wildlife conservation, but on which the protection of wild life from hunting is deemed important. All such reserves stand in recognition of the principle that protection of wildlife from predation by man is important to its long-term survival. Nevertheless, it is now generally recognized that complete protection of wildlife from hunting is only rarely essential to its conservation, and it is further recognized that protection from hunting alone is inadequate to maintain wildlife populations if equal attention is not given to the provision of a suitable habitat for the wild species.

Wildlife management, as the scientific application of ecological knowledge to the conservation of wild animals, had its origin for the most part on private hunting grounds and estates in Europe and Asia, but was formalized as a scientific profession in the United States during the 1930s. The work of Aldo Leopold, who wrote the first textbook of *Game Management* was particularly noteworthy in this respect. Leopold recognized that it was possible to maintain healthy, productive populations of wild animals through proper management of the habitats that supported them. Such populations could yield an annual crop for human use without in any way being endangered by this process. Such a crop could be taken by those who wished to hunt as a sport, or it could equally well be cropped commercially for the production of furs, hides, meat or other animal products.

Essential to the practice of wildlife management is recognition that hunting must be restricted to the sustained yield that can be removed safely without impairing the survival of the breeding stock or interfering with the productivity of the animal population. Where such restrictions cannot be imposed, the continued hunting, trapping, snaring or otherwise capturing of wild animals will lead to depletion of populations and, in some cases, to extermination. Thus the successful practice of wildlife management as a form of conservation depends upon the availability of a trained body of people able to enforce protective laws and regulations and to carry out the necessary studies and management practices needed to maintain the wild animals and their habitat. Such a corps of wildlife experts has been generally available in North America, Europe, Australia and, to a lesser extent, in some of the African countries. Generally speaking, such personnel have not been available, nor have the means for training them been available, in most of the developing countries of the world; nor, unless funds and technical skills can be made available from outside, are such people likely to be trained and employed. Consequently, despite the existence of protective laws and regulations, the hunting and trapping of wildlife continues with very little check or control. As populations increase and demands on the land grow, the threat of extinction faces an increasing number of wild species. The problem is accentuated by the peculiar tastes and fashions of those in developed countries who provide the market for the skins of rare animals and the various other products obtained from a variety of endangered species.

The commercial exploitation of wildlife for meat, hides, furs, ivory, etc., can provide the economic basis for the careful management and protection of a wildlife resource. Unfortunately, it also provides encouragement for the depletion of that resource in countries where protective laws cannot be enforced. Although there is every reason to encourage the production and trade in wild animals where this can be controlled and will not endanger animal populations, there is equal reason to restrict trade in species that will be endangered by excessive cropping. To this end, an international convention leading to the control of trade in endangered or depleted species of wild animals was drafted and made available for consideration by governments in 1971. Whether or not such an international agreement will accomplish more than the national laws now in effect remains to be seen.

The new interest in conservation

In the 1960s and early 1970s there was the beginning of a new, broadly based interest in the conservation of the human environment. This was hardly a spontaneous development, but rather a necessary response to the increase in the extent and severity of environmental problems. It is difficult to identify any single events which reflect the growing concern about the environment; however, a few will serve to illustrate the process. Thus, in the Soviet Union, scientists in the early 1960s became concerned about the potential for pollution of Lake Baikal that would result from the development of industries in the watershed of this unique water body. More so than any fresh-water lake in the temperate regions of the Earth, Lake Baikal has a distinctive, endemic fish fauna with hundreds of species to be found nowhere else in a lake. An irreparable loss to science would occur were these to be depleted or destroyed as a result of pollution. Fortunately, appropriate action to prevent pollution of the lake was taken by the government agencies concerned and plans are under way to create a major national park in the Lake Baikal area. Similarly, scientific concern for the falling water levels of the Aral Sea and the Caspian Sea, as a result of major irrigation diversions in their watersheds, has led to scientific studies of the potential consequences and plans for appropriate remedial action.

In northern Europe those nations bordering the Baltic Sea were faced with a problem of growing pollution and fisheries depletion in that major water body. Pesticide and heavy-metal pollution was of great concern, but perhaps of equal consequence was growing eutrophication leading to serious oxygen depletion and hydrogen sulphide generation in the deeper areas of the sea. Sweden took an early lead in the initiation of actions needed to bring this situation under control.

In Great Britain an active programme of air-pollution control went into effect in London with the result that the once-notorious black fogs that had beset that city and caused many deaths ceased to occur. A similar programme directed toward clearing up pollution of the River Thames produced excellent results. By 1970 dozens of species of fish that had not been seen in the river for many years were once more beginning to swim past the city of London. A programme to control the use of persistent pesticides also produced results, and once-depleted populations of hawks, ospreys and falcons began to come back to the British countryside.

In Japan the city of Tokyo was faced with impending disaster from air pollution which had grown so bad that it was necessary to supply oxygen artificially to traffic policemen in the more congested areas of the city. Starting in the late 1960s a major programme for pollution control was initiated. Among the developing nations, India was one of the first to recognize that environmental problems were not solely the concern of the wealthy nations. The five-year plan which went into effect in 1970 devoted significant attention to the environmental consequences of proposed development activities. Increasing effort was put into bringing India's population growth under better control.

In the United States, serious attention to environmental problems at the federal level led to ever-growing extensions of government action and expenditure to restore the quality of the environment. Most noteworthy was the increase in public interest and in attention of the news media to environmental problems.

All of this interest in the environment at the national level resulted in the realization that many of the major problems could not be tackled without international cooperation. New strength was given to those international programmes intended to provide a stronger scientific base for rational use and conservation of the environment, including those to be described in the next chapter. Popular interest in any subject is necessarily fickle but, with the increasing severity of environmental problems, there seems little likelihood that interest in their solution will fade away. When and if the problems are solved, the need for conservation, as such, will have ceased to exist. It will have become a normal part of man's daily life. But such a goal seems far in the future. Up to now, conservation has been a holding action, and its few gains have been far outweighed by the spreading deterioration of the human environment.

Chapter 6
New Goals, New Decisions

Extending the boundaries of knowledge

There is no question that the highest priorities today must be given to action programmes – to putting into operation the knowledge we have already acquired about how the biosphere works and how man's activities are related to its proper functioning. Nevertheless, anyone who is working in the biological sciences knows full well that, once action programmes start, we will soon run out of the knowledge needed to operate them effectively. Our knowledge of how the biosphere functions is still limited. Our knowledge of any single ecosystem, even those most thoroughly studied, still shows major gaps. There is an outstanding need for a major international programme designed to fill in the information gaps that exist. We must find out more about the techniques and management procedures required to more effectively guarantee the long-term productivity and health of the human environment. These gaps are obvious in the natural sciences, but are even more pronounced in the social sciences. A few examples are illustrative.

We still have very little knowledge of the effects of various degrees of crowding upon human populations. There have been many studies of animal populations and all of them show distressing parallels to observed situations among human beings in densely populated areas. Thus, John Calhoun and others have experimented with populations of rats and mice, providing them with every necessity in the way of food, water, etc., but limiting them in space. The animals increase to a high density but, at an early point in crowding, they begin to exhibit many behavioural abnormalities. In human terms, they exhibit a breakdown in sexual mores; rape and murder become rampant, young do not survive for lack of normal maternal care, hooliganism on the one hand, and complete social withdrawal on the

other, are everywhere evident. The behaviour too closely
resembles that exhibited in the slums of any large human city.
Rodents have not learned about drugs, nor have these been
made available, but if they were, there seems little doubt that
these animals would be making full use of them. But rats and
mice are not people. Where are the studies of human behaviour
under varying degrees of crowding related to different types of
physical and social organization? From all indications, the very
few studies that have been reported are entirely inadequate to
answer the basic questions that the city planners and architects
should be asking. How high dare we pile people in the cities of
the future? How many human units do we crowd into a physical
space before the units cease to be human? Surely major studies
in every kind of city, in every cultural situation, need be made
before we go on building houses and urban communities that
may serve only to cause the breakdown of their inhabitants.
International organization is required to provide the leadership
and coordination needed in such studies.

At the other extreme, we do not have nearly enough information about the existing status of animal species in relation to
their future survival. The International Union for Conservation
of Nature (IUCN) has assumed responsibility for monitoring
the population levels of species considered to be rare or endangered, and publishes the Red Data Books on endangered
species of vertebrate animals. But with the limited money that
has been available for its operation and the lack of knowledge on
how best to monitor populations of the less obvious or more
elusive wild species, it is impossible to report with any accuracy
on any but the most obvious and best-studied wild species. We
know how many California condors and whooping cranes there
are because the United States spends thousands of dollars on
counting and protecting these species. We have only the most
sketchy information, however, about the present abundance of
jaguars, ocelots, and other large cats – animals that enter into the
international fur trade to a degree measured in millions of
dollars. When it comes to plant species and the hundreds of
thousands of species of invertebrate animals, we have little
knowledge at all about survival or abundance. A few groups are
reasonably well known – shells of value to collectors, pine trees,

some butterflies – but most are not known. In most cases, it is hardly practical to census or monitor them individually. It is practical, however, to institute a better system of monitoring for the ecosystems to which they belong. All existing arrangements are inadequate – we must do better. But this requires organization, research and, therefore, money.

As has been noted, the recognition of great gaps in biological knowledge led to the launching of the International Biological Programme in the mid-1960s. The plan was to produce a coordinated, international study of those basic biological questions most important to the survival of the natural ecosystems of the planet and of man as a functional part of those systems. Although the programme produced good results in some nations, in an international sense it has not lived up to early expectations. The coordinated efforts have not always occurred. In many nations, no funds were available for any form of activity. The reasons for these difficulties can be variously analysed, but one reason is apparent. The programme was launched by the International Council of Scientific Unions, a non-governmental organization to which scientific groups in various nations belong. It was not successful in obtaining a general commitment on the part of the governments of the world to provide the necessary support for its programme. From the beginning, it struggled along without adequate financing, and the fact that it has succeeded as well as it has reflects unusual devotion and effort on the part of those who subscribed to it. But research is not cheap, and we cannot hope in the future to acquire the knowledge that we need by half-hearted means. The need for a much more major international effort is apparent.

Technology and pollution

In 1971, a new steel mill was opened in Texas in an area next to a major wildlife refuge. From all early indications it was of a different kind than any which had been operated in the United States before. Its smoke stacks release virtually no fumes, only a mixture of warm air and carbon dioxide. It requires some 682 000 m³ of water in circulation per day to keep operating, but only 46 000 m³ are drawn from the local water supply. The rest is re-circulated through efficient filtration and cooling systems. Twenty-three thousand cubic metres are lost daily through evaporation, and an additional twenty-three thousand are discharged into the adjacent bayou. The water discharged is cleaner than the water obtained from the local water supply. Excess salts, sludge and waste have been removed, condensed and solidified and used for land fill. The cost of this mill is high. Ten per cent of the total cost of the construction was spent on pollution controls. The cost of building similar controls into an already established mill would of course be much higher. All of those concerned with the control of pollution have praised the company responsible for the building of a non-polluting mill. Now, the international economic realities must be faced. The products built from the steel produced by this mill must compete on a world market with the products from all other mills. If the other mills have spared themselves that extra 10 per cent cost, will they be able to drive the non-polluting mill out of the market? In the long run, of course, the necessary costs of pollution control will far exceed the 10 per cent spent on pollution prevention. But that is in the long run, and economics and politics are heavily concerned with the short-run pay-off. Nevertheless pollution is being controlled effectively in many areas. The air of London and of Pittsburgh is far cleaner today than it was two decades ago. Many fish now live in the Thames that could not have survived there ten years ago. Birds of prey are once again returning to the British countryside, although effective control of insect pests continues.

There is little doubt that if price were no object we could build non-polluting industries everywhere, and re-fashion cities so that the total waste output into the environment would be minimal. The technology of re-cycling has been worked out for many materials and no doubt can be for most others. The cost of re-cycling and the market for the end products, however, present formidable obstacles.

The motor car has been a major cause of urban pollution, so much so that when nearly every other major cause of pollution of the air has been brought under control – as they have in the city of Los Angeles, for example – the output from automobile exhausts is still sufficient to create a serious and dangerous air-pollution problem. Nevertheless in 1975 all new motor cars in the United States will be forced to comply with federal regulations which will cut the output of air pollutants to a minimal and probably acceptable level. These new cars will be more expensive, unless perhaps the decision is made to abandon conventional internal-combustion engines and turn to other less expensive varieties. It seems likely, however, that the new cars will not appeal to countries that prefer to risk dirty air for the economic advantage of cheap transport.

It is possible to tackle all pollution problems everywhere in much the same way. But being possible does not mean that it is economically probable. Instead it seems essential to establish priorities, to determine not only what levels of pollution exist and the nature of the sources of pollution, but what levels are environmentally acceptable – capable of being broken down and re-cycled by natural processes in the air, water and soil of the biosphere. We need much more information on the environmental effects of various pollutants. At present this is being slowly and painfully acquired for existing chemical substances. However, new chemicals are being produced and released into the environment far faster than the old can be studied.

As a first step towards answering these questions, we need monitoring systems that will tell us accurately what problems exist and how they are developing. Such systems could be built up from existing global networks of oceanic and atmospheric monitoring stations, arranging that additional measurements and samples be taken at these stations. Further stations would need to be established, however, in order to acquire the full range of measurements required. From such networks we could determine existing levels of pollutants in the air, water, soil and at various levels in biological food chains, and we could find out trends in levels of pollution. When 'hot spots' or likely trouble areas were determined, action could then be taken that would prevent or ameliorate dangers that would otherwise occur. Without such a programme we are likely to waste much time and effort on low-priority tasks, while ignoring potentially serious, but unpublicized, problems.

Controlling land use

It would no doubt be generally agreed that, since land in general, and productive land in particular, is now in short supply relative to human needs and aspirations, the use of land should be controlled and lands should be used for the highest purposes. Such use would be taken to include non-use for commercially productive purposes where such non-use represented the higher value – as in lands set aside for scientific reservations, ecological reserves, national parks, or perhaps for the maintenance of various kinds of human cultures.

It is generally recognized that no individual or group can own land in any permanent sense. Each person holds land in trust for generations to come, each has a responsibility towards future generations to pass land down in an unimpaired or improved condition, except where the most urgent human purposes demand otherwise. It should be further accepted, after any study of history, that even nations have a similar role and responsibility. Each is a trustee over land on which the future generations of mankind must depend. Each has an obligation not just to those who live within its boundaries today but to all humanity for all time.

Considering these realities, one would expect much greater attention to the achievement of proper land use than is evident at the present time. But the present reflects conditions and attitudes of the past. For most of man's time on Earth, land was in abundance relative to foreseeable human needs. Only during the past century has this condition changed.

It is quite apparent that many nations today are unable to effectively control the use of land by their people. For many technologically advanced countries the power to control land use does not rest in the national jurisdiction but at some lower political level – state or county, canton or commune, province or prefecture. At best the national government can advise and hope to influence the decisions that will be made. In many nations, regardless of where the power of control legally exists, in operation it exists nowhere. There is no legal authorization for urban immigrants to build shanty towns around the boundaries of the large tropical cities, or to erect structures in city parks – but they do it nonetheless. There is no authority for people to pillage resources or to develop farms on land that legally belongs to the entire nation, but they do so. The means of control are either not available, or are not publicly acceptable. Nevertheless, control must be established.

International organizations cannot interfere with land use within any of their member nations, but they are in a good position to advise governments, to provide new information, and to arrange for exchanges of information between governments. Furthermore, they are in the best situation to support the kind of studies that would be most useful to their member nations – for example, studies on ways to influence behaviour through effective educational systems designed to operate within particular cultures. If we could find a way of influencing pastoralists in favour of proper stocking within the carrying capacity of their pastures and ranges, we should have a way to save millions of hectares of land that are now being converted into desert. The ecological information is available. The sociological and anthropological information is not. It would be helpful to all nations if they had access to information about systems of land-use control that have worked well within any single nation – examples might well include an analysis of the successes and failures of the British Town and Country Planning Act, of the French *Aménagement du territoire*', or of American purchase-and-lease-back systems of land-use control. Such information is currently available only to specialists with access to well-stocked libraries. It should be available to governments everywhere.

The conservation of natural environments

In 1972, with considerable publicity, Yellowstone National Park marks its centenary and a Second World Congress on National Parks will be held as part of the celebration. In one century the national-park concept has made some progress. The idea has been accepted and many remarkable areas of wild scenery and wildlife have been protected. Yet, it would be sanguine indeed to pretend that progress has been satisfactory. Throughout the world, wild nature is still in full retreat and the gestures toward its conservation have been entirely inadequate. Those who seek to implement a realistic programme towards conservation of the natural landscape find in most nations that the leaders of government have thus far paid scant attention to their arguments. The realities of political or economic survival today appear far more compelling than any moves that will produce their pay-offs and benefits far in the future. Although most governments are willing to proclaim the existence of parks and reserves, few are able or willing to pay for them in terms of serious protection and management. Where skilled manpower and technical training are already in scarce supply, it seems futile to expect governments to devote a significant share of these human resources to any but the most pressing problems of survival. What then is the course to be followed by those who know that tomorrow's survival may well depend on actions that can only be taken today? There will be no hope if we wait much longer before establishing the protection needed for wild nature – it will not exist.

There seems no hope but through the launching of a major international programme aimed at the establishment and protection of a world-wide network of parks and reserves. This need has long been obvious and it has led to numerous steps towards such a programme. Unesco and FAO have both devoted considerable time and effort toward assisting nations in the planning, establishment, protection and management of national parks. The International Union for Conservation of Nature and Natural Resources has devoted much of its time and energy to this purpose. Nevertheless, all of these activities are on a minor scale compared to the major task that must be accomplished. Assistance must be given in the planning and definition of suitable parks and reserves. In particular, training must be provided for the personnel needed for the administration, management, protection and scientific studies required if such a system is to be made functional. Assistance must be given to create public awareness of the importance and value of the conservation of nature, with particular attention to the education of the young in the significance and usefulness of natural reserves. Funds must be provided to keep all of these activities going until such time as park systems can stand on their own and begin to yield economic returns to the nations in which they are situated. To start this kind of major international programme might cost as much as one large modern building. To keep it functioning for perpetuity would probably cost less than 1 per cent of the amount that nations now spend in preparations for waging war.

Human cultures and the quality of life

What do we do about the human environment? How best can we repair the damages that have been done and work towards making the biosphere a continuing home for humanity? It is more important that we ask the right questions at this stage than that we be ready with the quick answer. The questions are difficult. The answers may be a long time in coming. Action is needed now to keep things from getting worse, to keep the options open for the human race. But beyond that are the more difficult problems – what kind of a world do you want to live in and to hand down to your children, what sort of a future do you want to see for mankind? Not everyone will be interested in these questions. Some are concerned with the present and the immediate future and do not care to look too far ahead. The future of nations and of man is not their concern and they will not make it their concern. But it is the concern of governments. It is the business of those who direct the activities that will shape tomorrow's world to think beyond today's well-being and provide for tomorrow.

The question of human numbers is central to any concern about the environment. Obviously, if populations were to grow without limit, we should soon, in an historical time scale, run out of space and resources. How soon is not difficult to calculate if we assume constant rates of increase. With a world population doubling every forty years, there would be 7 000 000 000 people in 2010, 14 000 000 000 in 2050, and 112 000 000 000 by AD 2170 – two centuries from now. Not the most ardent believer in the powers of science and technology proposes that the world could support over one hundred thousand million people. The more optimistic settle for fifty, or thirty or, more realistically, seven thousand million, while realizing that the world is not truly supporting the 3 600 000 000 that inhabit it today. So we are faced with the reality that growth rates will decline, that growth will stop, and we recognize that this can come in only two ways – fewer must be born, or more must die. Not many will vote for the second way.

But the question of how many people the world can support is surely the wrong question. Nobody who has seriously thought about the problem believes that the goal of human existence is to see how much human biomass can be crowded into the biosphere. Nobody looks forward to experiencing the maximum degree of crowding that humanity can tolerate and still survive.

Both Unesco and FAO have examined the problem of human numbers and human goals and, in a joint report submitted to the Economic and Social Council of the United Nations, have stressed the following points:

There seems little disagreement that the scientific knowledge and the technology developed during the twentieth century hold out a hope for mankind that was never before justified. The prospect of meeting the needs of the existing world population for food, fibre and other essentials for living is brighter than could have been foreseen a few decades ago. The prospect of meeting the needs of an ever expanding world population, however, is as remote as it ever has been. The availability of scientific techniques does not guarantee their application to human problems. The existence of a technology capable of providing benefits to all mankind does not necessarily imply that it can or will be used to improve human welfare.

To a surprising degree we have allowed our technology to dictate the conditions under which we must live, rather than using it to create a better environment. Too often we have allowed our knowledge of what is technologically possible to influence our predictions of what is economically and socially probable. Thus we hold out false hopes of prosperity and abundance.

We have basically three choices or various modifications of these, to select from:

Minimum subsistence for maximum numbers

We can channel all available resources and use all available space to provide a subsistence existence for the maximum number of people that the Earth can keep alive. All of us would consciously reject this as a goal. In practice, however, humanity is behaving in many parts of the world as though this were indeed the goal. The hundreds of millions of people existing at the edge of famine in parts of Latin America, Africa and Asia are evidence enough. The ultimate consequence of pursuit of this goal is the recurrence of catastrophes.

High material standards for maximum numbers

We can channel all available resources and use all available space to provide a high material standard of living for the greatest number of people that can be accommodated at such a level, disregarding natural values in favour of an artificial existence. Although not stated as a goal, this is a consequence, perhaps unforeseen, of the 'religion of growth' that is widely accepted by industrialized, technological societies. In these, natural values are usually sacrificed in favour of increasing production of those commodities that enter the market place and show up in indices of economic gain. The daily life of a middle-class apartment dweller in downtown Tokyo or New York could exemplify the relatively high material standard of living, high level of consumption of economic goods and almost complete lack of contact with things that are not man made, that would characterize the continued pursuit of economic gain at the expense of natural values. Whether this could be achieved by everyone is, however, doubtful.

The results of continued emphasis upon ever increasing commodity production and population growth at the expense of the natural environment are predictable: increasing pollution of air, land and water; disruption of natural ecosystems to the point where productivity and life are threatened; a constant war to control pests and plagues; and, ultimately, perhaps the evolution of a different kind of human being, able to tolerate such an existence, with the elimination of all those who cannot. Continued emphasis on economic growth at the expense of natural values, together with the inequities in distribution of material goods that appear to accompany such a pursuit, could increase the probability of international conflict through growing competition for markets and raw materials to a degree that the continued existence of mankind would be in doubt.

Quality of living for optimum numbers

The human goal implied in the concept of rational use of the biosphere is one that would seek a combination of a high material standard of living with the retention of a maximum variety of natural and man-made environments, including protection of non-human species and the values of wild nature. In such an environment, there would be the retention of opportunities for a change of direction and for the creation of different ways of living, since all resources would not be channelled or utilized and an abundance of living space would be available. Attainment of this goal would be possible only for a human population held at a compatible level, perhaps a level that could be described as an optimum abundance of people. The actual numbers involved in such an optimum population cannot be described in general terms, since they will vary with nations, cultures and levels of technology. From an ecological point of view this concept corresponds with that of an optimum density for an animal species, one at which we seek to maintain those animals that we manage. From an ecological viewpoint also, this appears to be the only realistic goal for humanity, one in which the survival of free, psychologically whole individuals remains possible. Indeed, whether or not it is accepted as a goal, this orientation toward quality of life in place of quantity of people and economic production is the only chance for retaining permanency of human civilization with full opportunities for individuals to develop their human potentials.

The question of what constitutes an optimum human population for the biosphere is in some ways meaningless, since there is no such thing as a population level called an optimum. What is optimum for one purpose or one technology may be too high or low for another. Furthermore, population questions must be answered in relation to particular peoples, areas and cultures, and not purely in global terms. Nevertheless it is a useful question, for it forces us to consider certain other questions about the future of the biosphere that are in themselves of a global nature. Among these are the following:

Do we want to preserve on the planet truly extensive wild areas such as our forefathers had available to them only a century ago? This does not refer to the relatively small areas now being protected in national parks and reserves, but to the broader sweeps of wild landscape that permit the continuance of primitive ways of living in natural environments – areas such as are still to be found in Antarctica, in Greenland and some other areas of the polar and subpolar regions, that still exist in the basins of the Amazon and Orinoco, in the Sahara and in the deserts of western Australia, and very few other places. Many such areas have been changed and modified through the activities of technological mankind, but they could be protected from further abuse and thus restored. The decision to maintain or not to maintain such areas is a major national and international decision, since for many countries it cannot be decided positively without major international support. The decision to foreclose this option, to agree to the modification or 'civilization' of such areas is a grave one. At stake is not just the great array of non-human wild nature that is involved, but also the ways of life of primitive peoples who should at least have the option to continue with these ways if they so desire. Have we decided that they and their ways must go, that they must be 'assimilated' into the technological world?

The answer does not just affect primitive peoples that exist today, but entire primitive ways of life that might be continued or reinstated by other people in the future. There is evidence enough today of a desire among the children of technology to seek some older and more satisfying way of life. But primitive ways of life – hunting, fishing and food gathering – presume the continued existence of broad areas of space and wild animal life. Their continued existence can mean the maintenance of a dimension to human life that has always been present during man's tenure on Earth. Have we decided deliberately to close this down?

Presumably, people are too civilized today to behave as their ancestors have too often done. The American Indians and their old ways of life did not vanish from most areas of the American continents through a deliberate planned policy of genocide or 'culture-cide'. They vanished from many areas because of the lack of policy or the presence of too many conflicting policies toward their lands, their cultures and their survival. Admittedly it is always easier to avoid answers to such difficult questions as the survival of wild lands and primitive people by turning one's back on the problem until it disappears – the people, the animals, the land somehow have 'unaccountably' faded from the scene. But this is hardly a procedure to be recommended in AD 1972. Rather one must ask what answer is proposed for this particular problem and what procedure will be followed in implementing it?

If we consider the preservation of older and more simple ways of living as one of the options to be decided by the international community, a second one of equal importance involves the preservation of somewhat more complex ways of life: of nomadic pastoralism, of primitive agriculture and the rural village life associated with it. These are ways of life that still exist in many areas of the globe. Although nomadism has been opposed by many governments and major efforts up to and including armed warfare have been brought to bear on its elimination, it still persists, albeit somewhat modified, in the Sahara, in Arabia and in some other areas. Primitive peasant agriculture persists more widely, from the highlands of South-East Asia, through the tropics of most continents and into some areas of the technologically advanced countries of the temperate zone. These are ways of life of value to mankind, available options for groups that do not necessarily like the urbanized technological existence that now dominates the planet. We can allow them to disappear by default by failing to do anything to preserve them. We can deliberately decide to eliminate these options and to force all to participate in modern agriculture or some other 'acceptable' form of technologically oriented existence. Or we can deliberately decide to preserve these options by setting aside appropriate areas in which such pursuits can be followed without interference. The qualification 'without interference' is an essential provision. If such ways of life are not allowed adequate space and an appropriate environment, experience has shown that they will be modified and 'swamped out' by the now dominant forms of animal husbandry and commercialized farming.

What is asked of the international community is that the issue be faced and a decision be made on rational grounds, one that the decision-makers are then prepared to defend.

A United Nations scientific committee seems to have already

endorsed such a stand in the following statement: 'The resources of the entire world must be developed rationally – to the fullest extent possible with the means available. Mankind as a whole can progress only by efficiently utilizing all of the Earth's available natural resources, especially at a time when its population is growing at such a startling pace.'

If such a goal is accepted, then the possibility of maintaining the older, less efficient ways of living with the environment is dismissed in favour of the goal of *efficient* use of *all* the Earth's available resources. However, one can question whether this is to be an accepted policy and intention of the international community.

There are other ways of life that require space and protection if they are to be preserved, including many that are only now beginning to disappear: the way of life of the small farmer, owning his own land and caring for it in his own way; of the independent commercial fisherman with his own boat; of all the great variety of small entrepreneurs in city or country, each in his own way making or selling some type of goods. None of these ways are economically efficient. All are eliminated by the competition of mass enterprise, whether it be socialistic or capitalistic in orientation. Yet all represent optional ways of living for those who prefer to be less efficient, less wealthy, but reasonably independent of the pressures of mass society. The 'pre-modern' ways of living need serious care and consideration if they are to be preserved, and one needs to ponder the consequence of a decision, overt or concealed, to eliminate them.

Such questions have a bearing on the issue of optimum population and quality of living for each country. Optimum populations will necessarily be at a lower level if space and opportunity are to be provided for maintaining the range of options that some consider essential to a high quality of environment.

Each community, each region and each nation needs to face the issue of growth relative to quality of living in its own way. Different cultures will find their own solutions, but only if they ask the question. If the question is not asked and decisions are not firmly implemented, then by default the options will close, the different cultures themselves will disappear, ground under and homogenized in a one-world, mass culture. A one-world, mass culture may be what people will want, but the decision in favour of it should be deliberately taken.

The question of what constitutes 'quality of living' for different peoples of different cultures, and how it relates to the ecological realities of their existence, is one for which a thorough study is badly needed. One can only hope that such a study is launched, the issues faced, and some answers determined while the opportunities to implement these answers remain available.

Public information

In the 1970s people are becoming increasingly aware that there are serious environmental problems. At the same time they are highly confused about the nature of these problems, their magnitude and severity, and the means for doing something about them. Misinformation about the environment has been provided in almost equal ratio with truth. People have been alarmed about the wrong things, disturbed about false issues and misdirected in their well-intentioned efforts. A few examples may help.

An alarm was raised in the late 1960s that, because of the increased consumption of fossil fuel, the burning of which consumes oxygen and releases carbon dioxide, in combination with the decrease in the amount of natural green vegetation on land and the growing pollution of the seas, the biosphere was in danger of running out of oxygen. This is a logical assumption on the basis of certain categories of existing information, and certainly a subject that deserves some study. It was, however, hardly a subject for major public concern. In the 1970s, when all of the existing information was pooled, it became quite apparent that there was no foreseeable danger of an oxygen shortage on the planet – that this was a 'non-problem' hardly deserving of any concern in comparison with the many real problems that exist.

In the 1970s there is a great wave of public sentiment against the killing of seals – 'baby seals' it is pointed out – to provide skins for the fur trade. The campaign ignores the fact that all wild animals die, that none of them die in comfortable beds in a hospital and that, once a habitat is fully stocked with any species of animals, any excess must inevitably die. The fact that man kills and uses some of them should hardly be considered as more a matter for public concern than the fact that man kills and eats cattle or sheep. One may demand that such killing involve a minimum of brutality and suffering. There is reason for serious public concern when the killing of any animal exceeds its natural rate of increase and the species becomes endangered. But this is hardly the case with the populations of northern fur seals on the Pribiloff Islands or of harp seals in Canada, which have been the target of the public anti-killing appeal. Rather, the northern fur seals are one of the very few examples of a successfully managed marine resource, protected by international treaty, brought back from near extinction and now cropped on the basis of safe levels determined by careful scientific study. To cripple this international agreement because of misunderstanding of the biological facts of life would be a serious setback for the conservation of wild species. Yet it seems likely to occur.

Perhaps most serious of all has been the tendency to equate conservation with low levels of consumption of natural resources and with a total rejection of technology. Such a view of conservation, while it may have an appeal to those who have always had high levels of consumption and an oversupply of technology, has no appeal whatsoever to the underprivileged majority of mankind. Furthermore, it has very little to do with the goals of conservation which are properly directed toward achieving a high material standard of living based on rational use of the Earth's resources. There is no necessary conflict between conservation and technology or conservation and international development. Only with the aid of the highest technology can the goals of conservation now be achieved. Only through adequate attention to ecological knowledge and conservation values can the goals of economic development be achieved without serious and unwanted environmental disruption. Only through development can most of the world's people gain sufficient necessities to begin to appreciate the meaning of 'environmental quality'. Admittedly, all people everywhere cannot live in the reckless and wasteful ways that have characterized certain 'privileged' segments of humanity. But the question is, Who really wants to live that way when better choices are available?

The need to control technology and regulate development in order to guarantee all people a chance for permanent achievement of a satisfactory way of living in pleasing environments that can be sustained is recognized by all concerned with conservation. Obviously, it is of equal concern to all but the most short-sighted advocate of unrestricted development.

These examples suggest the need for a responsible international action to provide public information on the facts concerning the biosphere and its use, and for a public-information programme of a far-reaching nature that will build support for rational means for accomplishing the mutual goals of conservation and economic development.

Chapter 7
An International Programme

The establishment of the United Nations in 1945 and of its various specialized agencies in the years that followed indicated a willingness of governments to act in concert in efforts to solve major problems confronting the world. No one would pretend that the United Nations has functioned in the ways that the founders of this organization had hoped. Many have suggested ways in which the United Nations could be improved. Nearly all would agree however that if the United Nations system were to cease to exist an organization very similar to it would have to replace it immediately. The willingness to abdicate any significant degree of national sovereignty is not yet apparent among the nations of the world. The willingness of nations to work together toward certain common goals is apparent. One goal on which all nations can agree is the need for maintenance and improvement of the human environment. The United Nations and its specialized agencies represent a means by which governments can work together toward this goal.

The Biosphere Conference held in Unesco House in Paris in 1968 was a major step forward for the United Nations system. It represented the first major intergovernmental conference directed towards solving the problems of the biosphere in its totality. The principles and state of knowledge affecting the rational use and conservation of the resources of the biosphere were brought under critical review. On the basis of the information presented, the representatives of the nations which were in attendance agreed on a series of recommendations for future action. The problems discussed so far in this book were generally recognized and the need for all governments to join in a broad programme directed towards their solution was stressed. Unesco was called upon to take the lead in forming a scientific research programme that would carry on from where the International

Biological Programme left off, but with the full support of governments and of the United Nations specialized agencies.

Following the Biosphere Conference, work was commenced in Unesco which led to the drafting of a major, long-term research effort that became known as the Man and the Biosphere Programme (MAB). This was done, as all such things must be, through a series of consultations with the member states of Unesco, with the specialized agencies of the United Nations system and with concerned non-governmental, international agencies such as the International Union for Conservation of Nature and the International Council of Scientific Unions. As finally drafted, the Programme consisted of a series of projects related to four major areas of study:

Natural environments

If man is to continue to exist it is essential that he be provided with more adequate knowledge of the world's ecosystems, of their characteristics, structure, functioning, productivity and their limits of tolerance to disturbance. Through study of natural ecosystems he can best learn how to manage such complex entities in order to maintain a sustained and maximum degree of benefit, both from the remaining natural environments and from those that have been modified to suit human needs.

The projects in this area of study are therefore directed towards a description and inventory of the world's ecosystems, studies of their structure and functioning, and measures needed for their conservation.

Of particular interest in the latter category are projects directed towards the establishment of a coordinated world-wide network of national parks, biological reserves and other protected areas, and those directed towards the conservation of species of wild plants and animals. These would serve to safeguard the ecosystems and species to be studied in the other projects of this series.

Rural environments

The projects in this area of study are directed primarily towards environments used for agriculture, pastoralism and forestry. Such rural systems are distinguished from natural systems in that, when monoculture replaces diversity, intrinsic instability takes over from stability. The broad objective of these projects is therefore to help achieve ecosystem stability concurrent with management of such domesticated ecosystems. The projects proposed would examine the various processes by which man has attempted to enhance the productivity of ecosystems – the use of fire, fertilizers, introduced species, modifications of plant cover, and various intensities and methods of cultivation. They examine the ways in which people have adapted to various extreme or highly fragile environments, and the ecological limitations of these environments for human use. They also examine and compare various parameters of natural and man-influenced environments in their relation to the sustained productivity which man hopes to achieve.

Urban–industrial environments

The rapid growth and spread of modern urban–industrial technology, combined with the growth of human populations, has had major effects upon all areas of the globe. Within urbanized or urbanizing areas the problems of human existence have grown complex and baffling, since the environments created by man often seem detrimental to his continued well-being.

Much harm has been done by failure to consider in advance the full range of ecological and sociological consequences resulting from a continued course of human action, or from a major restructuring of the environment through the use of engineering or technological skills.

Difficulties have also resulted from a failure of human societies both to give serious consideration to the kind of environment in which they aspire to exist, the level of quality they hope to obtain, and to relate their activities in ways that will truly achieve these goals. The projects in this area of study are intended to provide criteria by which man may direct his technology in the future and avoid those side effects which are harmful to his existence.

Pollution and related phenomena

A number of pollutants, like the components of the atmosphere, recognize no political or regional boundaries, and any research or monitoring activity must obviously be undertaken on a global scale. In order to establish the facts about pollution and pollutants, quantitative assessments will be made of the production, distribution and accumulation of many types of environmental pollution or degradation, together with evaluations of their effects on man, agricultural and other economic crops, and on wild vegetation and animals. The ultimate objective of much of the research proposed in this area of study is to assist in the establishment of environmental criteria, guides and standards for the biosphere, to facilitate the maintenance and amelioration of man's material and aesthetic existence.

The draft of the Man and the Biosphere Programme was presented to the Unesco General Conference in October, 1970. The debate among the delegates from various nations reflected a lively interest in the topic and virtually no disagreement on the importance of the Programme. Although there was considerable discussion of the means by which such a programme should be implemented, its cost, timing and the agencies to be involved in carrying it out, the Programme was approved and the necessary budget needed to set up direction and coordination within the Unesco Secretariat was voted.

An International Coordinating Council made up of scientific representatives from twenty-five nations, along with observers from appropriate United Nations and non-governmental agencies, was established and given the task of formulating detailed projects and priorities for the Programme. It was agreed that the Council would coordinate the Programme and Unesco would provide its Secretariat, but the actual work would be carried out by nations under the coordination of National Committees established for this purpose. For particular projects, appropriate inter-governmental or non-governmental agencies would be assigned responsibilities for leadership and coordination. Thus, certain projects would fall within the competence of the Food and Agriculture Organization (FAO), others would be particularly related to the work of the International Union for Conservation of Nature and Natural Resources (IUCN), and still others might be best coordinated through the International Council of Scientific Unions (ICSU).

There is of course a long way to travel between the setting up of a major international programme and the actual accomplishment of work in the field. The importance of MAB cannot be overstressed – none of the major areas proposed for study can safely be neglected. Yet, at the same time, there must be a willingness on the part of nations to meet the expenses and effort involved. Scientific talent must be mobilized, and this takes time and money. Public enthusiasm must be sufficient to guarantee that this time and money will be made available.

The wheels of international machinery grind slowly. Five years will have passed between the Unesco General Conference of 1966, which first called for the convening of the Biosphere Conference, and the convening of the first meeting of the International Coordinating Council for the Man and the Biosphere Programme. During those years pollution grew worse, deserts spread, species moved closer to extinction and the human condition generally grew more unstable. But nations hesitate to commit themselves to new programmes, they must be convinced not only by rational argument but by the actual pressures exerted by a deteriorating environment. There seems to be no way of speeding these commitments except by public demand from within each nation and this requires a growing level of public awareness not only of the problem but of the means and machinery required for its solution.

In 1968, another step was taken by the United Nations in the meeting of its General Assembly. At the prompting of Sweden, the General Assembly agreed to the urgency of the problems affecting the human environment and called for a major United Nations conference on this subject. Subsequently it was decided to convene this 'Conference on the Human Environment' in Stockholm in 1972. Unlike the Biosphere Conference, which was a meeting of scientific delegates, the Stockholm conference will attempt to attract the highest possible level of governmental representation. The aim will be the initiation of concrete action, through treaties, conventions and other inter-governmental means, directed toward the actual solution of environmental problems. One can hope, further, that, as a result of the interest spurred by the Stockholm conference, support will be forthcoming for those programmes (such as MAB) already in existence. There are obvious international actions to be taken at Stockholm on the basis of existing knowledge, but much that needs doing must be based on the kinds of information that the Man and the Biosphere programme can provide.

In 1972, it is difficult to be hopeful about the prospects for man and the biosphere he now controls. There is always the danger that the nations of the world, like the infamous Kilkenny cats of Ireland, will keep clawing, scratching and biting at each other until there is nothing left of them but their tails. But ecologist Sir Frank Fraser Darling summed up the situation as well as anyone can in addressing the Biosphere Conference: 'Ecologists can scarcely afford to be optimists. But an absolute pessimist is a defeatist, and that is no good either. We see there need not be complete disaster and if our eyes were open wide enough, world wide, we could do much toward rehabilitation.' Later, in his Reith Lectures on the BBC, which did much to awaken the people of Great Britain to the environmental problems they face, he stated: 'For myself, I have no doubts of the intention and earnestness of nations to act well and dress well, as it were. But time, it seems to me, is not on our side. Even by the 1972 Conference there will be over one hundred and fifty million more mouths in the world, all of whom will be demanding of technology, "Give us more... "; not just food, but more of everything.' How long will it take to start the research programme to gather the facts on which our actions to save man and his biosphere must be based? How long will it take for governments to take these actions and for public opinion to accept them? The answer seems to be related to the old question 'How much do you want to pay?' For the cost of perhaps ten space rockets we might take meaningful steps to save the colour of our blue planet, to save the life-sustaining layer of the biosphere. Can we afford it?

Further Reading

A. Arking and R. Jastrow, 'Exploring the mysteries of the planets and the cosmos', in A. B. Bronwell (ed.), *Science and Technology in the World of the Future*, Interscience, 1970, pp. 33–60.

J. B. Bresler, *Human Ecology*, Addison-Wesley, 1966.

A. B. Bronwell, *Science and Technology in the World of the Future*, Interscience, 1970.

A. Bryant, *Set in a Silver Sea*, Doubleday, 1968.

C. D. Darlington, *The Evolution of Man and Society*, Allen & Unwin; Simon & Schuster, 1969.

R. F. Dasmann, *Wildlife Biology*, Wiley, 1965.

R. F. Dasmann, *Environmental Conservation*, 2nd edn, Wiley, 1968.

R. F. Dasmann, A People at War with their Land, *National Geographic*. (In press).

K. Davis, 'The urbanization of the human population', in *Cities*, a *Scientific American* book, Knopf, 1965, pp. 3–24, Penguin, 1967, pp. 11–32.

J. Dorst, *Avant que nature meure*, Neuchâtel, Delachaux & Niestlé, 1965.

C. A. Doxiadis, 'Cities of the future', in A. B. Bronwell (ed.), *Science and Technology in the World of the Future*, Interscience, 1970, pp. 61–94.

C. Elton, *Animal Ecology*, Sidgwick & Jackson, 1927.

F. Engels, *The Condition of the Working-Class in England in 1844*, Leipzig, 1845, Translation, London, 1887: a new English translation by W. O. Henderson and W. H. Chaloner, *The Condition of the Working Class*, Blackwell, 1958.

F. Fraser Darling, *Wilderness and Plenty*, BBC Publications; Houghton Mifflin, 1970; Pan, 1971.

F. Fraser Darling and R. F. Dasmann, 'The ecosystem view of human society', *Impact of Science on Society*, vol. 19, Unesco, pp. 109–22.

C. Glacken, *Traces on the Rhodian Shore*, University of California Press, 1967.

E. H. Graham, *Natural Principles of Land Use*, Oxford University Press, 1944.

P. Hall, *The World Cities*, World University Library, McGraw-Hill; Weidenfeld & Nicholson, 1966.

G. Hardin, 'Not peace, but ecology', in *Diversity and Stability in Ecological Systems* (Brookhaven Symposia in Biology, 22), National Technical Information Service, 1969, pp. 151–161.

IUCN, *Red Data Book*, vols. 1–5, Morges, IUCN, 1966–70.

IUCN, *United Nations List of National Parks and Equivalent Reserves*, IUCN Publications, New Series, no. 15, Morges, 1971.

G. P. Kalinin and V. D. Bykov, 'The world's water resources, present and future', *Impact of Science on Society*, vol. 19(2), Unesco, 1969, pp. 135–50.

A. Leopold, *Game Management*, Scribner's, 1933.

T. Malthus, *An Essay on the Principle of Population as it affects the Future Improvement of Society, with Remarks on the Speculations of Mr. Godwin, M. Condorcet, and other writers*, St. Paul's, 1798; Macmillan, 1926; Penguin, 1970.

W. Marx, *The Frail Ocean*, Ballantine, 1967.

J. Milton and M. T. Farrar (eds.), *The Careless Technology*, Doubleday. (In press)

L. Mumford, *The City in History*, New York, Harcourt, Brace & World, 1961; Penguin, 1966.

E. P. Odum, *Fundamentals of Ecology*, 3rd edn, W. B. Saunders, 1971.

A. I. Oparin, *Origin of Life*, Macmillan, 1938; Dover, 1953.

Population Reference Bureau, *World Population Data Sheet*, Population Reference Bureau, 1970.

D. Ratcliffe, 'Decrease in egg shell weight in certain birds of prey', *Nature*, vol. 215, 1967, pp. 208–10.

R. Rudd, *Pesticides and the Living Landscape*, University of Wisconsin Press, 1964; Faber, 1965.

C. O. Sauer, *Agricultural Origins and Dispersal*, American Geographical Society, 1952; M. I. T. Press, 1969.

C. O. Sauer, 'Study of critical environmental problems', in *Man's Impact on the Global Environment*, M.I.T. Press, 1970.

K. P. Shea, 'Unwanted Harvest', *Environment*, vol. 11, 1969, pp. 12–16, 28–31.

G. Sjoberg, 'The origin and evolution of cities', in *Cities*, a *Scientific American* book, Knopf, 1965, pp. 25–39; Penguin, 1967, pp. 33–48.

A. G. Tansley, 'The use and abuse of vegetational concepts and terms', *Ecology*, vol. 16, 1935, pp. 284–307.

J. O'G. and J. H. Ruzicka, 'Organo chlorine pesticides in antarctica', *Nature*, vol. 215, 1967, pp. 346–8.

W. L. Thomas (ed.), *Man's Role in Changing the Face of the Earth*, University of Chicago Press, 1956.

Time, 'Environment: the clean machine', *Time*, 17 May 1971, p. 46.

S. L. Udall, *The Quiet Crisis*, Holt, Rinehart & Winston, 1963.

Unesco–F A O, *Conservation and Rational Use of the Environment*, United Nations Economic and Social Council, United Nations (E/4458), 1968.

Unesco, *Proposals for a Long-Term Intergovernmental and Interdisciplinary Programme on Man and the Biosphere*, Doc. 16/C/78, Sixteenth General Conference, Unesco, 1970.

Unesco, *Use and Conservation of the Biosphere*, Unesco (National Resources Research, X), 1970.

United Nations, *Demographic Yearbook*, 1960, Department of Economic and Social Affairs, United Nations, 1960.

United Nations, *Problems of the Human Environment*, Report of the Secretary General, United Nations, 1969.

United Nations, *Natural Resources of Developing Countries*: *Investigating Development and Rational Utilization*, Department of Economic and Social Affairs, United Nations, 1970.

S. M. Wadham and G. L. Wood, *Land Utilization in Australia*, Melbourne University Press, 1939.

A. Woodbury, *Principles of General Ecology*, McGraw-Hill, 1954.

Acknowledgements for Illustrations

Photographs
  1 Rapho, Paris. Photograph by Michel Bossé.
  2 Black Star, New York. Photograph by Marvin Hardee.
  3 Rapho, Paris. Photograph by Georges Gerster.
  4 United States Information Service.
  5 Western Ways Features. Photograph by Tab Nichols.
  6 Hoa-Qui, Paris.
  7 Rapho, Paris. Photograph by Emile Schulthess.
  8 Unesco. Photograph by Paul Almasy.
  9 United States Information Service.
 10 Gamma, Paris.
 11 Rapho, Paris. Photograph by de Sazo.
 12 United States Information Service.
 13 United States Information Service.
 14 Hoa-Qui, Paris.
 15 Rapho, Paris. Photograph by Christian Zuber.

Figures
  1 Unesco.
  2 Unesco.
  3 US Department of the Interior.
  4 Unesco.
  5 *Le Courrier de la Nature*.